Technology: A Second Level Course

ANALOGUE AND DIGITAL ELECTRONICS

BLOCK 4 TRANSISTORS AND BASIC CIRCUITS

Prepared by the Course Team

The Open University

THE ANALOGUE AND DIGITAL ELECTRONICS COURSE TEAM

AUTHORS

Dr David Crecraft

Dr David Gorham

Roger Loxton

Dr Mike Meade *Course Team Chairman*

Dr John Newbury

Dr Phil Picton

Dr Ed da Silva

Prof John Sparkes

COURSE SUPPORT STAFF

Steve Best *Graphic Artist*

Ian Every *Academic Computing Service*

John Garne *Academic Computing Service*

Ruth Hall *Designer*

Neville Hillyer *Project Officer*

Dr Mavis Hamilton *Course Manager*

Roger Harris *Course Manager*

Andy Reilly *Editor*

Karen Shipp *Academic Computing Service*

The Open University
Walton Hall, Milton Keynes, MK7 6AA

First published 1990. Reprinted with corrections 1994. Reprinted 1996.

Copyright © 1990 The Open University
ISBN: 0 7492 60173

Designed by the Graphic Design Studio of the Open University.

Printed in Great Britain by Hobbs the Printers Ltd, Southampton.

This text forms part of an Open University course.

For general availability of supporting material referred to in this text please write to: Open University Educational Enterprises Limited, 12 Cofferidge Close, Stony Stratford, Milton Keynes, MK11 1BY, Great Britain.

Further information on Open University courses may be obtained from the Admissions Office, The Open University, PO Box 48, Walton Hall, Milton Keynes, MK7 6AB.

Edition 1.3

PART 1 p–n JUNCTIONS AND TRANSISTORS

CONTENTS

STUDY GUIDE

As explained in the Study Guide for Block 4 as a whole, this part of Block 4 is structured in such a way that you can *begin* by studying just the *descriptions* of the devices' properties, only returning to study the explanations later when you are more familiar with what has to be explained. So each section is divided into a Part A and a Part B. Part A contains both descriptions and explanations, whilst Part B simply summarises the main points of the section. The statements in Part B are the essentials of what you need to know in order to understand Parts 2 and 3 of this block, though of course the statements themselves are not explained in Part B. The summary in **bold print**, which comes at the beginning of each Part B, is all you need to know to use the devices in circuits, and so forms a useful reminder of device properties to refer to in the later blocks of the course.

There are therefore two alternative routes through this text:

(i) If, to begin with, you prefer to read a *summary* of the explanations as well as a *description* of device properties, you should follow the 'B route', and learn Part B of each section.

(ii) But if you prefer to understand things as you go, you should follow the 'AB route' (i.e. the complete text) in the order presented.

(Note that the **bold route** on its own is enough for later blocks, but is not enough for Parts 2 and 3 of this block.)

Of course you are expected to study the whole of the text before you leave the block, but students differ as to their learning styles and you may be one of those who prefer to get the facts clear before trying to grasp explanations of them. If so, you should begin with the 'B route'. If, however, you find it difficult to learn bare facts without the explanations that make them make sense, you should follow the 'AB route' and work your way through the text in the usual way doing the SAQs when you come to them. Either approach can be equally satisfactory, so you can choose which you prefer, but don't forget that you should have studied the whole of the text before starting Block 5.

You will not be using your computer with this part of the block, but you will need a scientific calculator.

AIMS

The aims of this first part of Block 4 are:

1 To describe and briefly explain the main properties of silicon.

2 To describe the d.c. characteristics of diodes and of the principal types of transistors.

3 To explain, largely qualitatively, the d.c. characteristics of silicon rectifiers and transistors in terms of the properties of silicon and the structure of the devices.

4 To explain a simple application of these d.c. characteristics, namely the design of d.c. current sources using transistors.

OBJECTIVES

The objectives you should achieve in studying this text are set out below.

GENERAL OBJECTIVES

In order that the above aims can be achieved it is necessary that you learn the meanings of a number of new terms, namely:

As regards silicon

Acceptors, donors, doping, acceptor density, donor density
Carriers, majority carriers, minority carriers
Carrier density, minority carrier density, majority carrier density
Drift current, drift velocity, mobility
Equilibrium carrier density, intrinsic carrier density
Generation and recombination of hole–electron pairs
Holes
Infinite recombination surface, ohmic contact
n-type and p-type silicon

As regards p–n junction diodes

Diffusion, diffusion current, minority carrier density gradients
Forward and reverse bias of a p–n junction
Metallurgical junction, the transition region of a p–n junction
Saturation current I_S of a p–n junction

As regards bipolar transistors

Base, collector and emitter regions and terminals
Common base and common emitter connections of a transistor
Current gains α and β of a bipolar transistor
Current mirrors and d.c. sources
d.c. characteristics of a bipolar transistor
Early effect, Early voltage
Epitaxial transistors, planar process
Output resistance of a transistor and of a current mirror

As regards field-effect transistors (JFETs and MOSFETs)

Channel length modulation factor, output resistance
Depletion-mode and enhancement-mode MOSFETs
Drain, gate, source and substrate of a MOSFET or JFET
Gain factor β of a MOSFET
Input resistance and capacitance of a FET
Linear and saturation regions of operation
n-channel and p-channel MOSFETs and JFETs
Pinch-off point, pinch-off voltage of a JFET;
Threshold voltage of a MOSFET

SPECIFIC OBJECTIVES

When you have completed your study of this text, you should be able to:

1   Calculate the conductivity of silicon under different conditions of doping and temperature.

2   Calculate the current through diodes and transistors under different conditions of bias.

3   Explain, in terms of the properties of silicon and the structure of the devices, the principles of operation of diodes and transistors.

4   Describe different ways of constructing current sources and compare their performance.

1 INTRODUCTION

In the previous two blocks you have been studying how to arrange circuit elements like amplifiers and logic gates so that they will achieve a specified overall function. In Block 1 you studied some of the *passive* circuit elements used in such circuits. This part of Block 4 continues with the study of circuit elements but concentrates mainly on *active* ones, namely the various types of transistor.

As explained in the study guide there are two possible approaches to the study of this text. You can either concentrate to begin with only on the *descriptions* of the device characteristics in the B part of each section, and return to study the *explanations* later, or you can study the whole text in which each section depends to some extent on the explanations in the previous one.

Now, as you probably already know, most transistors nowadays are made from silicon, so this material is the first topic to be discussed. However transistor operation depends above all on the properties of p–n junctions, which are junctions between two different types of silicon. p–n junctions are also at the heart of most rectifier diodes these days, so they are the subject of the Section 3. Finally, two quite different generic types of transistor are described and explained in Sections 4 and 5; namely the bipolar transistors and the field-effect transistors. There are several variants of each of these basic types, as you will soon discover, which will also be described and explained.

2 SILICON

(A) DESCRIPTION AND EXPLANATION OF SILICON PROPERTIES

2.1 RESISTIVITY AND CONDUCTIVITY

When a conducting material such as copper or silicon is formed into a specific shape, such as a wire or rod, its resistance depends on its shape as well as on a property of the material called its **resistivity** or **conductivity**. Resistivity is simply the reciprocal of conductivity.

If a material has a conductivity of σ, and a rod or wire of length L and cross-sectional area A is made from the material, as shown in Figure 1(a), it will have a resistance R given by

$$R = \frac{\text{resistivity} \times \text{length}}{\text{cross-section area}} = \frac{1}{\sigma} \times \frac{L}{A} = \frac{L}{\sigma A}. \tag{1}$$

Resistivity has the units ohm-metres (Ω m), so conductivity has the units siemens per metre (S m^{-1}, which is the same as Ω^{-1} m^{-1}). As illustrated in Figure 1(b), resistivity can be thought of, rather unrealistically, as the resistance between opposite faces of a 1 metre cube of the material (i.e. a sample whose length $L = 1$ metre and whose cross-sectional area $A = 1$ square metre).

SAQ 1	(i) If the conductivity of copper is 5.8×10^7 S m^{-1}, what is the resistance of a length of copper wire 100 m long and 0.5 mm in diameter?

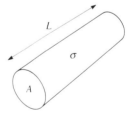

(a) $R = \dfrac{L}{\sigma A}$

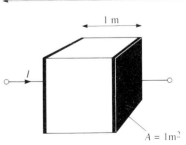

(b) $R = \dfrac{1}{\sigma}$

FIGURE 1 the relationship between resistance, R, and conductivity σ; (a) in a rod or wire of cross-sectional area A; (b) in a 1 metre cube, whose resistance = 1/conductivity

(ii) A wafer of silicon of length 0.5 cm is 0.5 mm thick and 0.2 cm across. If the resistance between its ends is $100\,\Omega$ what is the conductivity of the silicon?

2.2 SILICON CRYSTALS

The conductivity of silicon can be altered by a process known as **doping**. This consists of adding elements from Group III or Group V of the Periodic Table of Elements to pure silicon. It is the essential first step towards making silicon devices. You can picture what happens during doping as follows.

Silicon (Si) is a member of Group IV of the Periodic Table, which means that each atom of silicon has four outer electrons available to form bonds with other atoms. The result is that the atoms of pure silicon tend to form diamond-like, tetrahedral crystals, as illustrated in Figure 2(a), in which each silicon atom is surrounded by four other silicon atoms. All four outer electrons of each atom are involved in these chemical bonds—with two electrons involved in each bond—so that most of them are fixed in the crystal lattice and cannot move, as indicated diagrammatically in Figure 2(b). So the crystal is a poor conductor of electricity. (Below about 200 K *all* the outer electrons are fixed in the lattice, so the material is an insulator.)

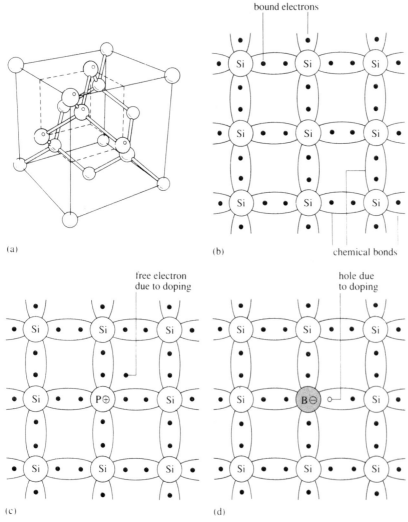

FIGURE 2 The structure of silicon crystals: (a) the tetrahedral structure of pure silicon crystal with the inter-atomic bonds shown as rods between the atoms; (b) a diagrammatic representation of the structure showing the presence of one electron from each silicon atom in each of the bonds; (c) n-type material with a donor atom replacing an occasional silicon atom and releasing a mobile electron into the lattice; (d) p-type material with an acceptor atom replacing an occasional silicon atom and creating a mobile 'hole' in the crystal structure

When, however, an occasional silicon atom is replaced by a Group V atom, such as phosphorus (P), which has five outer electrons and is called a **donor** atom, only four of these electrons are tied down in the bonds. This leaves one electron per atom free to move and conduct electricity, as indicated in Figure 2(c). So, to a first approximation, there are as many free electrons as there are donor atoms. Silicon doped with donors is called **n-type silicon** because electric currents in it are carried mainly by (negatively charged) electrons. Similarly, when a Group III element such as boron (B), which has only three outer electrons and is called an **acceptor** atom, is added, a gap or **hole** is created in one of the bonds, as indicated in Figure 2(d). It turns out that this hole is also free to move and conduct electricity. That is, when an electron from a neighbouring bond moves into a hole in the bond, the hole has in effect moved in the opposite direction. *A hole behaves like a mobile positive charge in the crystal and can be thought of in those terms.* Silicon doped with acceptors is therefore called **p-type silicon**.

It should be appreciated that all the other electrons which form part of the silicon or dopant atoms remain fixed in the crystal lattice and so can take no part in any conduction processes. Also it is worth noting that only tiny amounts of doping material are involved. The density of silicon atoms in a silicon crystal is about 10^{29} m^{-3}, whereas the typical doping densities are between 10^{20} and 10^{24} m^{-3}, so the concentration of donors or acceptors may well be less than 1 part per million! It is remarkable that such a small change in the constitution of silicon can bring about these profound changes in electrical behaviour.

2.3 CONDUCTION IN SILICON

Conduction in n-type and p-type material is illustrated in Figure 3, in which only the mobile **carriers** are shown. (The term 'carriers' is used to refer to either mobile electrons or mobile holes, or to both.) If a voltage is applied to p-type material, the mobile holes will drift towards the more negative terminal, as shown in Figure 3(a). And if a voltage is applied to n-type material, electrons drift towards the more positive terminal, as shown in Figure 3(b). Both forms of carrier flow constitute an electric current. Note that holes flow in the direction of conventional current, but electrons flow in the opposite direction.

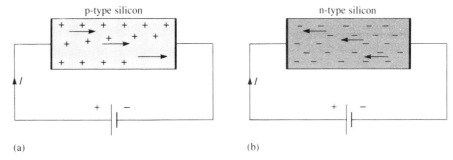

(a)   (b)

FIGURE 3 Current flow in semiconductor materials: (a) the drift of mobile holes in p-type material when a voltage is applied; (b) the drift of electrons in the opposite direction in n-type material.

It is mainly because semiconductors can conduct electricity either by means of holes or by means of electrons, or sometimes by both together, that semiconductor devices possess their extraordinary properties.

Although the mobile electrons and holes in a semiconductor at room temperature are in fact always in a state of rapid random motion due to their inherent thermal energy, it is only the average 'drift' of the carriers in the direction of the electric field that concerns us here. The electric current that it represents is called a **drift current**. The average velocity at which the carriers move in the electric field is called their **drift velocity** and is proportional to the applied electric field. So it is convenient to define a parameter called **mobility** μ which is the drift velocity of the carriers per unit electric field. The units of mobility are metres-per-second per volt-per-metre, or $m^2\,V^{-1}\,s^{-1}$.

> The mobility μ_n of electrons in silicon is $0.15\,m^2\,V^{-1}\,s^{-1}$.
> The mobility μ_p of holes in silicon is $0.045\,m^2\,V^{-1}\,s^{-1}$.

You don't need to remember these values, but it is worth noting that electrons are normally over three times as mobile as holes.

Now, remembering that conductivity can be thought of as the conductance of a 1 metre cube, and that current is the rate of flow of charge through a cross-section of the conductor, the conductivity σ is given by

$$\sigma = (\text{carriers per } m^3) \times (\text{mobility of carriers})$$
$$\times (\text{charge per carrier}).$$

The charge per carrier is the charge of an electron, called the 'electronic charge' $q = 1.6 \times 10^{-19}\,C$. The number of donors per cubic metre is called the **donor density** N_d, and so this is also the density of mobile electrons released into the silicon material. So for a sample of n-type silicon doped to a density of N_d the conductivity σ_n is given by

$$\sigma_n = N_d \times \mu_n \times q. \tag{2}$$

Since mobility and electronic charge have fixed values, equation (2) gives

$$\sigma_n = N_d \times 0.15 \times 1.6 \times 10^{-19} = 2.4 \times 10^{-20} \times N_d.$$

Similarly, for p-type materal

$$\sigma_p = N_a \times \mu_p \times q \tag{3}$$

or $\qquad \sigma_p = 0.72 \times 10^{-20} \times N_a.$

For example, the conductivity of n-type silicon doped with donors to a density of $10^{22}\,m^{-3}$ is, from equation (2), $240\,S\,m^{-1}$.

SAQ 2	(a) What do you suppose happens to the conductivity of silicon: (i) when you dope a sample of silicon material with equal numbers of donors and acceptors; (ii) when you dope the sample with unequal numbers of both kinds of donor?

(b) A sample of silicon is doped with phosphorus to a density of $10^{21}\,m^{-3}$ as well as by boron to a density of $5 \times 10^{20}\,m^{-3}$. What will be the conductivity of the silicon sample?

2.4 GENERATION AND RECOMBINATION

So much for the effect of doping on silicon. I now want to consider pure silicon again and explain the effect of temperature change upon it; in particular why it is that the conductivity of pure silicon increases with increase in temperature, whilst that of many metals, such as copper, *decreases* with temperature.

Briefly, the reason is that the heat energy associated with temperature causes **hole–electron pairs** to be generated in the semiconductor. That is, some of the electrons acquire sufficient thermal energy to enable them to escape from their bonds and become free to conduct electricity, leaving holes behind that can also conduct electricity. The hotter the material the greater the average energy of the electrons and the greater the rate at which these hole–electron pairs are generated. However, there is bit more to it than that.

If the thermal generation of hole–electron pairs were the only process going on, the silicon crystal would soon fill up with carriers and would become as good a conductor as a metal. In fact there is a complementary process at work which is continually getting rid of the hole–electron pairs as they are generated, but not quite immediately. This further process is

These values for the mobilities are only approximate. Mobility falls off as the density of carriers increases. At a constant density of carriers, mobilities decrease as the temperature rises. To make matters simple we shall assume that μ_n and μ_p are constant at these values at all carrier densities at room temperature (300 K).

called **recombination**. If there are holes and electrons present in a silicon crystal at the same time they will recombine after an average time called their **lifetime**. Recombination refers to the process whereby electrons fall back into the holes in the bonds, thus removing free holes and free electrons in equal numbers simultaneously. Each electron that recombines becomes fixed in the lattice once again—where the hole was—and is no longer free to move, until it once again acquires sufficient energy to escape from its bond. When recombination occurs there is a release of energy which is equal to the minimum energy required to generate a hole–electron pair. This energy release either causes heating or, in some semiconductors, the emission of light. The red light-emitting diodes, used for example in some hand calculators or on the panels of hi-fi sets, derive their light from the recombination of hole–electron pairs in specially made semiconductors.

With both generation and recombination taking place together, an equilibrium density of holes and electrons is created. A balance occurs between the rate at which hole–electron pairs are being created and the rate at which they are recombining. In pure silicon at 25°C this balance results in there being an equilibrium density of 1.5×10^{16} holes and electrons per cubic metre. This quantity is called the **intrinsic carrier density**, n_i. (Note that this density amounts to only about 1 electron per million million atoms in the silicon!)

Now the rate of generation depends on *temperature*—the hotter the material the greater the generation rate. The rate of recombination however depends on *carrier densities*. This is because recombination depends on the probability of holes and electrons meeting within the crystal. So the magnitude of n_i increases with temperature. That is, more carriers must be present if the recombination rate is to match the increased generation rate and produce an equilibrium density of carriers. It turns out that n_i increases at a rate of about 8 per cent per degree centigrade.

SAQ 3

(a) What is the conductivity of intrinsic silicon at a temperature of 25°C?

(b) What is the conductivity of intrinsic silicon at 100°C assuming that the conductivity increases at a rate of 8 per cent per degree centigrade?

Now when silicon is doped, the effects of recombination and generation lead to an equilibrium state in which the hole and electron densities are not equal. If the equilibrium density of holes is p_0 and if the equilibrium density of electrons is n_0 it turns out that for all practical doping densities,

$$p_0 n_0 = n_i^2. \tag{4}$$

So, for example, in an n-type sample n_0 will be much bigger than n_i due to the doping, with the result that the equilibrium hole density p_0 must be much *less* than n_i. Similarly if the hole density is increased by doping with acceptors, the equilibrium electron density must decrease. More generally, if we call the doped-in carriers the **majority carriers** and the others the **minority carriers**, *doping increases the equilibrium density of the majority carriers and decreases the equilibrium density of minority carriers*.

Since the majority carrier densities are approximately equal to N_d or N_a, the minority carrier densities are given by

$$p_0 = \frac{n_i^2}{N_d} \text{ in n-type Si}; \quad n_0 = \frac{n_i^2}{N_a} \text{ in p-type Si}. \tag{5}$$

For example, if the donor density is 10^{23} m^{-3}, so that there are about 10^{23} m^{-3} free electrons in the crystal, then the density of holes in equilibrium is given by

$$p_0 = \frac{n_i^2}{10^{23}} = \frac{2.25 \times 10^{32}}{10^{23}} = 2.25 \times 10^9 \text{ m}^{-3}.$$

Note that this density is much less than the intrinsic density at room temperature and many orders of magnitude less than the donor density.

As the temperature increases the minority carrier density increases rapidly as already explained, so there is a temperature at which n_i becomes comparable with the density of doped-in carriers—either N_d or N_a. As you will see later, p–n junctions depend for their operation on there being a clear difference between the p-region and the n-region, so this temperature—when n_i is comparable with either N_a or N_d—sets an upper limit to the useful temperature range of such devices. In practice this temperature is around 150–200°C for silicon devices, though it depends on the doping densities. We will not be considering the behaviour of silicon at such temperatures in this course.

The *speed* at which equilibrium is achieved, once the equilibrium densities of holes or electrons have been disturbed, can vary a good deal between specimens of silicon. The actual rates of generation and recombination depend on the quality of the crystal as well as on the temperature. Any impurities and any defects in the regular lattice structure of the single crystal—including the surface of the crystal—provide **recombination** and **generation centres** which assist both processes. They provide stepping stones, so to speak, which enable the electrons to escape from the bonds or return to them more easily. Indeed, if the recombination rate is to be kept low enough to give minority carriers a useful lifetime within a silicon crystal, extreme care has to be taken in producing pure, fault-free single crystals before the doping and other processes begin. The fact that imperfections in the crystal affect both recombination and generation *equally* means that the equilibrium intrinsic carrier density is the same (1.5×10^{16} m^{-3} at 298 K) in all silicon crystals, however badly they are made. What is altered, however, is the *rate* at which equilibrium is restored once it has been disturbed, As you will soon see, the operation of semiconductor devices depends on the densities of minority carriers being changed from their equilibrium values in one way or another—by light or by injection through a p–n junction for example. If these changes do not have time to have their intended effect before the minority carriers recombine with the majority carriers, the devices will not work: rectifiers won't rectify, amplifiers won't amplify, and so on. This is why high-quality single crystals are required for the production of silicon diodes and transistors.

Note however what happens at the metal contacts that are made to semiconductors. If you refer back again to Figure 3(a) you will see that something rather surprising is taking place at the ends of the p-type bar. The current in the copper connecting wires is of course carried by electrons; but in the p-type bar it is carried by holes. These two types of carriers flow in opposite directions when they are carrying the same current. Evidently at the right-hand end of Figure 3(a) the holes and electrons are, therefore, converging on the end contact and immediately recombining there. Similarly, at the left-hand end, hole–electron pairs are generated as rapidly as the current demands—the holes moving away towards the right and the electrons going to the left. Metal contacts that have this property of, in effect, generating or recombining hole–electron pairs as required are called **ohmic contacts** to distinguish them from rectifying contacts. Such contacts can, for example, be made by pressing gold wire into the silicon surface to form a cold weld. They are also sometimes called **infinite recombination surfaces** to indicate that recombination and generation will take place as rapidly as the circuit requires.

These ohmic contacts have an important property; they *hold the carrier densities in the semiconductor material at their equilibrium values,*. That is, however much the carrier densities are disturbed locally, for example by applying a voltage to a nearby p–n junction, they are kept to p_0 and n_0, as determined by the doping level, at the semiconductor surface where the contact is made. In other words the *recombination in effect takes place so rapidly that no change from equilibrium values at the surface contact can persist for a significant time.*

Finally in this section it is worth noting that silicon is not the only material from which p–n junctions can be made: germanium and gallium arsenide are also used in making useful devices. Germanium was the first material to be used for making transistors when they were invented in about 1948, but has now been almost wholly superseded by silicon, partly because the upper temperature limit of typical germanium devices is only about 90°C and partly because of the much greater reliability of silicon devices. Gallium arsenide is used when extra performance is needed at high frequencies and high temperatures, but it is much more difficult to process than silicon so it only replaces silicon for specialised purposes. Silicon will surely remain the basic material of most of electronics for many years yet.

(B) SILICON AND ITS PROPERTIES SUMMARIZED

1   Silicon is a brittle, shiny, crystalline material that has a conductivity midway between that of metals and insulators.

2   Pure silicon in equilibrium contains equal densities of holes and electrons, called the intrinsic density n_i. $n_i = 1.5 \times 10^{16}$ m^{-3} at 25°C and increases by about 8 per cent per degree.

3   The overall conductivity of a semiconductor is the conductivity due to holes plus the conductivity due to electrons.

4   The conductivity of silicon can be altered by doping it with particular kinds of impurities. If a Group III element, called an 'acceptor', is added (e.g. boron, gallium, indium or aluminium) p-type silicon is formed in which current is normally carried by holes. If a Group V element, called a 'donor' is added (e.g. arsenic, antimony or phosphorus) n-type silicon is formed.

Note: Typical doping densities are in the range of 10^{20} to 10^{24} m^{-3}.

5   The equilibrium densities of holes and electrons result from a balance occurring between the rate of generation of hole–electron pairs, which depends on temperature, and their rate of recombination, which depends on their densities. If p_0 is the equilibrium hole density and if n_0 is the equilibrium electron density then $p_0 n_0 = n_i^2$. Doping with donors increases n_0 and therefore decreases p_0, whilst doping with acceptors does the reverse, the product always equalling n_i^2.

6   The conductivity of n-type silicon σ_n is $N_d \times q \times \mu_n$ (2)

whilst that of p-type silicon σ_p is $N_a \times q \times \mu_p$,   (3)

where N_a and N_d are the acceptor and donor doping densities

electronic charge $q = 1.6 \times 10^{-19}$ C

electron mobility $\mu_n = 0.15$ m^2 V^{-1} s^{-1}

and hole mobility $\mu_p = 0.045$ m^2 V^{-1} s^{-1}.

7   If both types of donor are present the effective doping density is the difference between the two densities, the larger density dominating.

8   The rates of hole–electron pair generation and recombination are affected equally by imperfections in the crystal. Thus to achieve long average lifetimes of carrier densities in *excess* of equilibrium values, very good quality crystals of silicon are needed.

9　At ohmic contacts (e.g. a cold weld between gold and silicon) the rates of recombination and generation are in effect almost infinite, so that the densities of holes and electrons cannot be significantly disturbed from their equilibrium values however much current flows.

10　Some typical conductivities (in S m^{-1}) at about 25°C

copper	5.8×10^7
n-type Si ($N_d = 10^{22}$ m^{-3})	240
p-type Si ($N_a = 10^{22}$ m^{-3})	72
polythene	10^{-20}

3　THE p–n JUNCTION

(A) DESCRIPTION AND EXPLANATION OF p–n JUNCTION PROPERTIES

3.1 STRUCTURE OF p–n JUNCTION DIODES

A **p–n junction** is a junction, within a single crystal, between p-type silicon and n-type silicon. In other words the doping changes from primarily acceptors to primarily donors at a surface within the crystal.

The usual way to produce a p–n junction is as follows. A thin slice or wafer of silicon, as illustrated in Figure 4(a), is first cut from an ingot of silicon crystal that has been produced in a special furnace. This starting wafer is doped during the growth of the single crystal by adding either donors or acceptors to the molten silicon so that, as it crystallises, the silicon becomes either n-type or p-type. This wafer forms the **substrate** for perhaps thousands of p–n junctions. Then by *diffusing* acceptors, at high temperature, into an n-type substrate, or by *diffusing* donors into a p-type substrate, as illustrated in the cross-section diagrams of Figures 4(b) and (c), single p–n junction are formed.

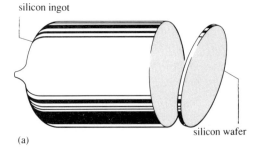

(a)

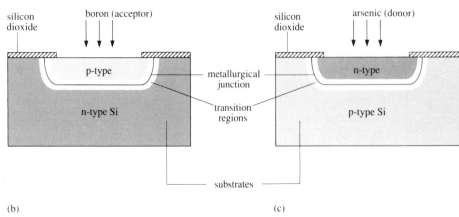

(b)　　　　　　　　　　　　　　(c)

FIGURE 4　Forming a p–n junction in silicon: (a) cutting a thin wafer from a single crystal ingot of silicon to form the substrate for many p–n junctions; (b) cross-section showing the diffusion of boron through a 'window' in the oxide layer that covers the surface of an n-type substrate to form a single p–n junction; or (c) diffusing arsenic into a p-type substrate. The extent of the transition region on either side of the metallurgical junction is shown by the white area

Diffusion is the name given to the way in which a crowd of particles in random motion tends on average to move from regions in which their density is high to regions in which it is low. This net movement occurs simply because there are more particles available to move away from high density areas than from low density ones, so the densities tend to even out. Perfume for example spreads through a room, even when there are no draughts, by this process of diffusion. In the same way donors or acceptors will diffuse into solid silicon if a high concentration of them is placed next to the silicon surface in an oven at a high temperature. The rate of penetration of the donors or acceptors depends on their concentration at the surface and on the temperature.

The diffusion process is used in the fabrication of silicon p–n junctions as follows. First a film of silicon dioxide is grown over the surface of the silicon by heating it in an atmosphere of oxygen. Then a hole—or 'window'—is etched in the oxide layer, as indicated by the gap in the oxide in Figures 4(b) and 4(c). When the silicon is heated to over $1000°C$ in an atmosphere of the dopant, the donor or acceptor atoms diffuse into the silicon through this 'window' in the oxide. The density of the dopant atoms decreases with distance from the surface; both depth and density can be accurately controlled by controlling the temperature and time of the process. If the starting material is n-type and the diffused material is an acceptor such as boron, as in Figure 4(b), the surface is converted to p-type silicon because the acceptor density there exceeds the original donor density. But deeper down the donors still dominate, so a boundary is formed between the p-region and the n-region at which the dominant dopant changes from donors to acceptors. This boundary is called the **metallurgical junction**. The p–n junction as a whole extends a little on either side of the metallurgical junction, to form the so-called **transition region** whose properties are at the heart of p–n junction and transistor action.

Figure 4(c) illustrates the complementary process of diffusing donors into a p-type substrate in order to convert the top surface to n-type. Again a p–n junction is formed at the interface.

With both forms of p–n junction, if wires are now attached to the two regions it will be found that it is possible to pass current through the device in one direction, but hardly at all in the opposite direction. The device is therefore a **p–n junction rectifier***. A typical d.c. characteristic, and the circuit for measuring it, are shown in Figure 5. In Figure 5(b) the current is almost zero with the voltage applied in one direction—called the *reverse*

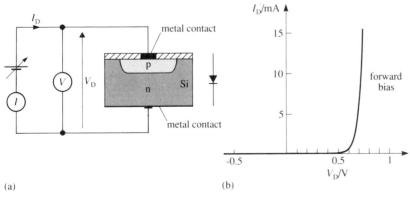

(a) (b)

FIGURE 5 The p–n junction diode: (a) a cross-sectional diagram of a silicon p–n junction, its graphical symbol and a circuit for measuring its d.c. characteristic; (b) a typical p–n junction d.c. characteristic

---

*p–n junction rectifiers are commonly called 'junction diodes', although strictly speaking this term refers to any 2-terminal device that contains a p–n junction. Such devices as Zener diodes, photo-diodes, light-emitting diodes, etc., are 2-terminal devices that contain p–n junctions, but they are not discussed in this course.

direction, but it increases exponentially when the applied voltage is in the *forward* direction. (*The forward direction is when the p-region is more positive than the n-region.*) This reverse direction of applied voltage is often called **reverse bias**. Similarly, the forward direction of applied voltage is usually called **forward bias**.

The rectifying action of a p–n junction depends, as explained in a moment, upon the majority carriers in one region being able to pass right through the transition region into the other region when the junction is forward biased, as indicated in Figure 6(a). It also depends upon there being almost zero flow of electrons and holes when the junction is reverse biased, as indicated in Figure 6(b); only the few thermally generated minority carriers from each region being able to pass through it; the majority carriers of course are driven away from the junction by the applied voltage.

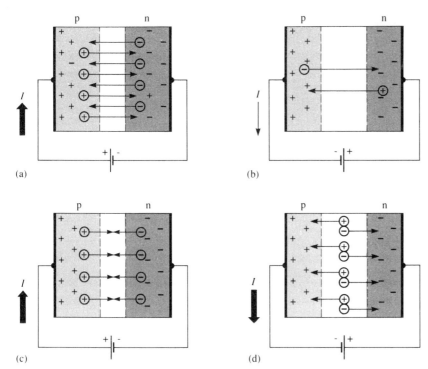

FIGURE 6 Carrier flow in a p–n junction: (a) forward bias when there is negligible recombination—the majority carriers pass through the transition region to become minority carriers on the other side of the junction; (b) reverse bias when there is negligible recombination—only the very few minority carriers are drawn through the junction, giving a tiny reverse current; (c) and (d) forward and reverse bias when there is heavy recombination and generation in the transition region. Under forward bias (c) the majority carriers recombine, and under reverse bias (d) hole-electron pairs are produced, giving a large reverse current

If the transition region of the p–n junction were full of crystal imperfections (e.g. if you were simply to press two crystals together) heavy recombination and generation would occur in the junction, as indicated in Figures 6(c) and 6(d). Under forward bias holes and electrons would converge on the junction and recombine, as in Figure 6(c), rather than pass through the transition region as required. Similarly, if such a junction were reverse biased, hole–electron pairs would be continuously generated at the crystal imperfections within the junction, giving rise to a large reverse current, as illustrated in Figure 6(d). So a current of similar magnitude would flow through the junction in either direction of bias. The device would be more like an ohmic contact than a rectifier because of the heavy generation and recombination occurring within it.

The transition region around the metallurgical junction is very sensitive to external pollution. If it is exposed to oxygen or water vapour, for example, its rectifying properties are severely affected as a result of the crystal imperfections caused by these impurities. For this reason it is important that the places where the p–n junction comes to the surface of the silicon, as the donors or acceptors diffuse sideways as well as downwards, are covered with the protective layer of silicon dioxide shown in Figures 4(b) and 4(c). One of the main reasons why silicon is used for integrated circuit construction is the ease with which this protective oxide layer can be laid down during the fabrication process.

3.2 OPERATION OF p–n JUNCTIONS

The d.c. characteristic shown in Figure 5(b) can be briefly explained as follows. A more detailed explanation is to be found in many textbooks on semiconductor devices. The detailed theory shows that:

(i) **The applied voltage determines the minority carrier density just next to the transition region of the junction.** First recall that the *minority* carrier densities in each doped region on either side of the junction are very small. (Doping increases the equilibrium *majority* carrier density, but decreases the equilibrium *minority* carrier density, as explained in Section 2.4.) The dashed lines in Figure 7(a) and (c) show the equilibrium minority carrier densities: p_{n0} of holes in each n-region and n_{p0} of electrons in each p-region.*

Figure 7(a) indicates how forward biasing *increases* the minority carrier densities next to the junction; Figure 7(b) shows that zero bias leaves the densities unaffected, whilst reverse biasing the junction *reduces* the minority carrier densities in both regions just next to the junction, as shown in Figure 7(c). All three diagrams also indicate that the ohmic metal contacts at the two ends of the piece of silicon hold the minority carrier densities at their equilibrium values, as explained in Section 2.4. And since the minority carriers diffuse away from regions of higher density towards regions of lower density, they distribute themselves through the regions on either side of the p–n junction as indicated by the density profiles shown in Figures 7(a) and (c). It is easy to show that if the region is uniformly doped, and if there is negligible recombination in either region, each density profile of minority carriers is linear as indicated in the figures.

The factor by which the minority carrier densities change at the edges of the transition region is $\exp(K V_D)$, where $1/K \approx 25$ mV. Thus, for example, a forward bias V_D, as in Figure 7(a), of 0.1 V, which is $4 \times 1/K$, produces a minority carrier density change just next to the transition region, of $e^4 \approx 55$. So the hole density on the n-region side of the junction becomes $55 p_{n0}$, and the electron density on the p-region side becomes $55 n_{p0}$. Thus significant density gradients are produced in both regions. Indeed the densities increase by a factor of about 55 for each further 0.1 V applied, so they can become very large compared with the equilibrium value with a bias of only a few tenths of a volt.

On the other hand, if the applied voltage is a *reverse* bias (e.g. $V_D = -0.1$ V) as in Figure 7(c), the minority carrier densities are reduced to $p_{n0}/55$ and $n_{p0}/55$. These densities are almost negligible when compared with the equilibrium densities, so the reverse bias gradients cannot increase much more even if the reverse bias voltage is increased further.

 SAQ 4   The n-region of a p–n junction is doped with phosphorus to a density of 10^{22} atoms m^{-3}. The junction is forward biased by 0.6 V. What is the density of holes in the n-region just next to the transition region at 25°C?

(ii) The second result that can be derived from theory is: **the current that flows through an isolated p–n junction is determined by the density gradient of *minority* carriers in the two regions on either side of the junction.** Consider

In these symbols, p_{n0} and n_{p0}, the first subscript indicates the region involved and the 0 subscript implies the presence of equilibrium conditions. Thus p_{n0} refers to the equilibrium hole density in the n-region.

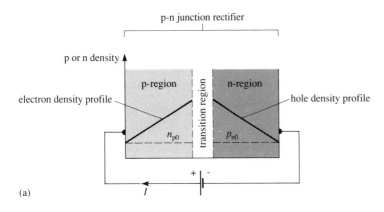

(a)

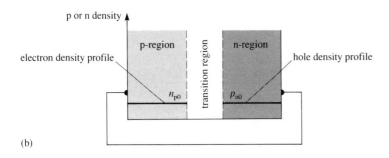

(b)

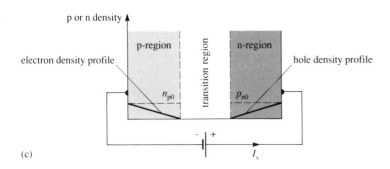

(c)

FIGURE 7 Diagrams showing minority carrier densities, p or n, within the regions on either side of a p–n junction. The circuit indicates the direction of bias. (a) Forward bias increases the minority carrier density just next to the junction; (b) zero bias leaves the densities at their equilibrium values; (c) reverse bias reduces the densities next to the junction. The currents are proportional to the minority carrier *density gradients*. The densities are held at their equilibrium values by the end contacts

forward bias first—Figure 7(a). Because the density of minority carriers next to the transition region is greater than at the metallic contacts there is diffusion of both kinds of carriers away from the transition region: the electrons in the p-region flow towards the left down their density gradient, whilst the holes in the n-region flow towards the right. Both constitute a conventional current in the direction shown. The total current through the diode consists of the sum of these two currents.

It is possible to derive the equation for this current as follows:

For any given size of a p-region, the electron density gradient in it is proportional to $(n_{p0} \exp(K V_D) - n_{p0})$, so, in the p-region,

$$\text{electron current is proportional to } n_{p0} \times (e^{K V_D} - 1) \tag{6}$$

Similarly, in a particular n-region,

$$\text{hole current is proportional to } p_{n0} \times (e^{K V_D} - 1) \tag{7}$$

Therefore

$$\text{total current } I_D \text{ is proportional to } (n_{p0} + p_{n0})(e^{K V_D} - 1).$$

Rewriting this equation, including an appropriate proportionality factor, gives the characteristic p–n junction equation, namely

$$I_D = I_S(e^{K V_D} - 1). \tag{8}$$

I_S is called the **saturation current** and is proportional to the sum of the minority carrier densities $n_{p0} + p_{n0}$. It is also a function of the sizes of the p-type and n-type regions on either side of the p–n junction—the larger the region the smaller the gradients for a given bias voltage.

☐ To produce low values of saturation current, should the regions be heavily doped or lightly doped?

■ Heavy doping of silicon increases the majority carrier density but decreases the minority carrier densities p_{n0} and n_{p0}. So heavy doping produces small values of I_S.

When reverse bias is applied, as in Figure 7(c), the gradients are in the opposite direction and so cause a reverse current. The same arguments apply, so equation (8) applies to both forward and reverse bias conditions.

In Figure 7(b) there are no gradients, so there are no currents.

| SAQ 5 | Calculate values of I_D (shown in Figure 5(b)) for a diode for which $I_S = 10^{-14}$ A at the following values of applied voltage V_D: (i) 0.7 V; (ii) 0.65 V; (iii) 0.6 V; (iv) 0.4 V; (v) -0.1 V; (vi) -5.0 V. |

It is worth remembering a few simple conclusions that follow from equation (8).

● At any current level an increase in V_D of 0.1 V results in a current increase by a *factor* of about 55.

● A current increase by a factor of 10 is produced by a voltage increase of about 57 mV.

● Current is doubled by a voltage increase of about 17 mV.

| SAQ 6 | In many silicon diodes a more realistic value of K is 35 V^{-1}. If the value of I_S is 10^{-13} A, what forward voltage would produce a forward current of 1 mA in such a p–n junction? What current would be produced by a forward voltage of 0.4 V? |

For most practical purposes when designing circuits you will find that it is sufficiently accurate to assume that:

(a) if a significant current is to flow through the junction, a forward voltage of between about 0.6 V to 0.7 V must be applied to it, and

(b) that with a forward bias of less than about 0.4 V, or with a reverse bias that is not so large as to cause the diode to break down, the current flowing is likely to be quite negligible.

It should be noted, however, that there is a limit to the amount of reverse voltage that can be applied to a p–n junction. Beyond a certain point the junction breaks down, which means that it starts to conduct a large current and is likely to overheat. This breakdown is not necessarily destructive, except as a result of overheating, but such voltages should be avoided. The manufacturers' data sheets specify maximum allowable currents and voltages.

Now, as already explained, the total current through the junction is the sum of the hole current and the electron current. In transistors it turns out that the *ratio* of these two components of the p–n junction current is an important consideration. For regions of equal size, it is clear from equations (6) and (7) that the hole current is proportional to p_{n0} and the electron current is proportional to n_{p0}. Therefore

$$\frac{\text{electron current}}{\text{hole current}} = \frac{n_{p0}}{p_{n0}}. \tag{9}$$

But n_{p0} is inversely proportional to the acceptor density in the p-region just as p_{n0} is inversely proportional to the donor density in the n-region, so

$$\frac{\text{electron current}}{\text{hole current}} \approx \frac{\text{donor density in n-region}}{\text{acceptor density in p-region}}. \tag{10}$$

Thus the current ratio is simply the doping ratio. The situation when the n-region of a p–n junction is more lightly doped than the p-region is illustrated in Figure 8.

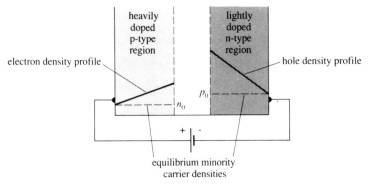

FIGURE 8 An asymmetrical junction with a much steeper hole density gradient in the lightly-doped n-region than the electron density gradient in the heavily-doped p-region. Therefore the hole current is greater than the electron current

One further feature of a p–n junction which does not affect its d.c. properties, but has important consequences for transistors, concerns the nature of the transition region. It turns out that the transition region of a p–n junction is almost depleted of carriers of either type; those that pass through it do so so rapidly that they do not significantly increase the equilibrium density within it. The consequence of this is that a p–n junction not only behaves like a rectifier, but also behaves like a capacitor; the two regions on either side behave like conducting plates and the transition region behaves like a dielectric. Indeed under reverse bias, when the d.c. current is very small, p–n junctions form quite good capacitors. This feature of a p–n junction is of importance when considering the high-frequency properties of transistors (see Block 7).

In addition, however, the width of the transition region depends on the voltage applied to it, as indicated in both Figures 6 and 7: the larger the reverse voltage the greater the width. This variation of transition region width means that the junction capacitances referred to in the previous paragraph are nonlinear. It also has important consequences for the d.c. behaviour of transistors, as explained in Section 4.

(B) p–n JUNCTIONS AND THEIR PROPERTIES SUMMARIZED

1   p–n junctions are made from a single crystal of silicon in which there is a transition from p-type material to n-type material at the metallurgical junction within the crystal. A 'transition region', which is largely depleted of carriers, extends a little way on either side of the metallurgical junction. A p–n junction therefore behaves rather like a capacitor with the silicon in the transition region acting as the dielectric.

2   In addition, however, a p–n junction forms a highly nonlinear resistor, with a typical characteristic like that of Figure 5(b). In other words, a p–n junction is a rectifier. Its characteristic obeys the equation

$$I_D = I_S(e^{KV_D} - 1) \tag{8}$$

where

- **I_D is the diode current and V_D is the bias voltage applied to the diode. V_D is positive for forward bias, which is when the p-region is more positive than the n-region. I_D is positive for forward current (i.e. flowing from the p-region towards the n-region).**

- **I_S is called the saturation current, and for small diodes is typically about 10^{-13} to 10^{-15} A. Evidently $I_D = -I_S$ when V_D is more negative than about -0.1 V.**

- K is a constant equal to about 40 V^{-1} (or $1/K = 25 \text{ mV}$) at room temperature.

  Note: The curve of Figure 5(a) is also the graph of equation (8) for a p–n junction in which $I_S = 10^{-14} \text{ A}$.

3 The symbol for a rectifier is shown in Figure 5(b). The arrow points in the direction of forward current.

4 The ratio of electron to hole current through a p–n junction is dependent mainly upon the donor/acceptor ratio in the two regions on either side of the p–n junction.

5 p–n junctions are made from high-quality crystals of silicon in order to minimise the effects of recombination and generation within the transition region, which would otherwise spoil the rectifying properties of the p–n junction.

6 The application of a voltage V_D to a p–n junction changes the minority carrier densities just next to the transition region by the factor $\exp(KV_D)$. Since the minority carrier densities are held at their equilibrium values at the ohmic contacts, minority carrier density gradients are created in each region which give rise to diffusion currents. The sum of these two currents is the total diode current I_D.

7 The application of a voltage to a p–n junction not only affects the current through it, it also affects the width of the transition region: the greater the reverse voltage the greater the width.

4 BIPOLAR TRANSISTORS

(A) DESCRIPTION AND EXPLANATION OF BIPOLAR TRANSISTOR PROPERTIES

4.1 STRUCTURE OF BIPOLAR TRANSISTORS

Silicon bipolar transistors are 3-terminal devices consisting of two p–n junctions which are formed back to back in a single crystal of silicon. They can be in either an n–p–n or a p–n–p configuration, as shown diagrammatically in Figures 9(a) and 9(b). The three regions of such a transistor are called **emitter**, **base** and **collector** as shown. The graphical symbols for the two versions of the device are also shown in these figures.

A cross-section of a typical n–p–n transistor structure is shown in Figure 9(c). Such a device is produced in much the same way as a p–n junction diode, except that there is one additional diffusion. Beginning with an n-type substrate a p-type region is diffused into it through a window in the protective oxide layer, as for a diode. But then a new oxide layer is grown over the surface, a new and smaller window is etched in it and some *donor* material is diffused into the p-type region, converting its surface layer back to n-type material again. This final diffusion produces the emitter region which, for reasons explained in a moment, has to be much more heavily doped than the base region, and is therefore labelled n⁺. This method of making transitors is called the **planar process**. The structure whose cross-section is shown in Figure 9(c) results. Note that the n-type collector consists of a thin layer of n-type material grown on top of a thicker substrate of n⁺ material. This n⁺ substrate is thick enough to provide mechanical strength, and its heavy doping ensures that the collector region has only a small resistance. This process of growing a lightly doped layer of single crystal silicon on top of a heavily doped substrate, as in the figure, with no break in the crystal structure, is called 'epitaxy'. Transistors made in this way are called **epitaxial transistors**. Typical dimensions and doping densities of a discrete planar epitaxial silicon n–p–n transistor are shown in Figure 9(d).

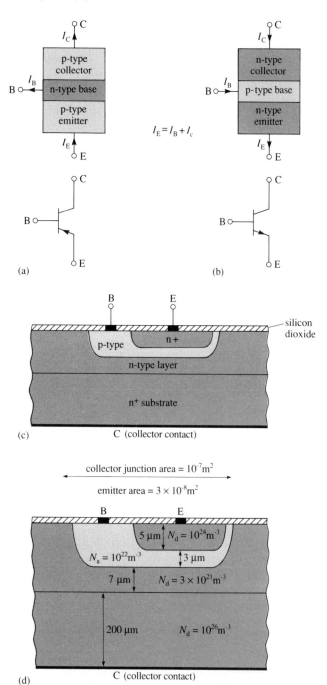

$$I_E = I_B + I_c$$

FIGURE 9 Bipolar transistors: (a) a diagrammatic representation of a p–n–p transistor, plus the standard graphical symbol for a p–n–p transistor; (b) the same for an n–p–n transistor; (c) a typical structure of a discrete n–p–n transistor; (d) some typical dimensions and doping densities of such an n–p–n transistor

☐   What is the effect on the emitter current of doping the n-type emitter region more heavily than the p-type base region?

■   It ensures that the emitter current flowing through the emitter junction will consist mainly of electrons. The hole current will be relatively small (less than 1 per cent in practice).

In a p–n–p transistor the p-type emitter region is doped more heavily than the n-type base region, so that the emitter current I_E consists almost entirely of holes flowing from the emitter into the base region.

When bipolar transistors are constructed in integrated circuits some additional features are included, but these need not concern us here.

4.2 OPERATION

The essence of transistor action is the following: by applying a small voltage between the *base* and *emitter* terminals of the device (0.65 V is found to be approximately the voltage required to forward-bias a silicon p–n junction) it is possible to cause a current to flow in the *collector* circuit, as indicated in Figure 10. The large arrow indicates the path of most of the current. The key to understanding bipolar transistor action lies in understanding why this happens; why the majority of the emitter current flows in the *collector* circuit rather than in the forward-biased *base–emitter* circuit where the battery is.

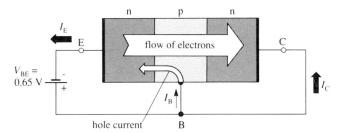

FIGURE 10 Carrier flow in an n–p–n transistor with zero collector-base voltage indicating how most of the emitter current consists of electrons flowing through to the collector even when the collector is connected directly to the base (i.e. $V_{CB} = 0$)

Normally a d.c. voltage source is also connected between collector and emitter in order to reverse bias the collector–base p–n junction and provide more power gain, but the collector current will flow as shown without it. So why does the current that enters the base region from the emitter preferentially leave the base region via the collector lead, rather than via the base lead, even when there is no battery in the collector-base lead?

To answer this question we need only look at the minority carrier gradient in the base region. Figure 11(a) shows the electron density profile in the p-type base region of an n–p–n transistor.

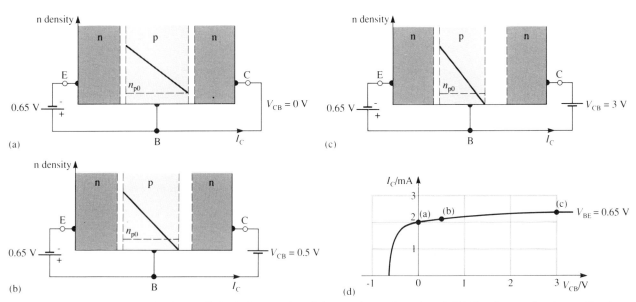

FIGURE 11 (a) The electron density profile in the base region of the transistor when $V_{CB} = 0$. (b) The electron density profile when $V_{CB} = 0.5$ V. (c) The profile when $V_{CB} = 3$ V; the narrowing of the base region caused by the widening of the transition region is called the Early effect. (d) The resulting d.c. characteristic curve of the transistor

☐   What is the effect on minority carrier densities just next to a p–n junction of applying a voltage V across the junction?

■   The minority carrier densities just next to the transition region are increased under forward bias, and decreased under reverse bias, by the factor $\exp(KV)$; see Section 3.2.

Suppose n_{p0} is the *equilibrium* density of electrons in the p-type base region. At the emitter end of the base region the density will be raised above n_{p0} by the forward bias applied to the emitter–base junction, whilst the zero voltage applied to the collector–base junction will hold the density at the collector end to the equilibrium value as shown. Evidently, as the electrons diffuse away from the high density near the emitter junction, a steep minority carrier gradient which spans the base region will be created between emitter and collector and will give rise to a current from emitter to collector. Note that the applied *voltages* ($V_{BE} \approx 0.65$ V and $V_{CB} = 0$ V) hold the densities at the two ends of the base region constant so that the emitter-collector current flows continuously.

If there is no recombination of hole–electron pairs in the base region, the electron gradient is the same right through the base region so that the emitter and collector *electron* currents will be equal. Note however that a small fraction of the emitter current—usually less than 1 per cent—flows from base to emitter in the form of a hole current, as indicated by the small arrow in Figure 10, so that the total emitter current I_E is slightly larger than the collector current I_C. (Remember that conventional current flows in the opposite direction to electron flow.)

☐   How is it arranged that the electron current from the emitter into the base is much larger than the hole current from the base into the emitter region?

■   By doping the emitter region much more heavily than the base region, as explained in Section 4.1 (see also equation (10)). See Figure 9(d) for typical doping densities.

The ratio of I_C/I_E is called α, the **common-base current gain**. The difference between the emitter and collector currents is the base current I_B (that is, $I_E = I_C + I_B$) and the ratio of the collector current to base current is called the **common-emitter current gain** β.

That is,

$$\beta = I_C/I_B. \tag{11}$$

If there is some recombination in the base region as the electrons diffuse across it, the collector current is decreased slightly by this loss of electrons and the base current is increased by the same amount. Despite these effects, however, β usually lies somewhere between 100 and 500. Thus despite the two processes that give rise to a base current the collector current—due to diffusion of electrons through the base region—remains dominant.

☐   Show that $\alpha = \beta/(1 + \beta)$ and that $\beta = \alpha/(1 - \alpha)$. If the range of values of β is 100 to 500 what is the range of α?

■   You know that $\alpha = I_C/I_E$ by definition. It is also evident that $I_E = I_C + I_B$ by Kirchhoff's current law, so $\alpha = I_C/(I_C + I_B)$. Dividing both the denominator and the numerator of the right-hand side of this equation by I_B, and remembering that $\beta = I_C/I_B$, leads to the equation $\alpha = \beta/(1 + \beta)$. This equation can be rewritten as $\alpha(1 + \beta) = \beta$. Rearranging gives $\alpha = \beta - \alpha\beta = \beta(1 - \alpha)$. Therefore $\beta = \alpha/(1 - \alpha)$. If $\beta = 100$, then $\alpha = 100/101 \approx 0.99$. If $\beta = 500$, $\alpha = 500/501 \approx 0.998$.

Figure 11(b) shows the same diagram as Figure 11(a), but with a low-voltage battery in the collector-to-base lead which gives a *small* reverse bias to the collector-base junction. The effect is to reduce the hole density at the edge of the collector junction to well below n_{p0}. This increases the electron density gradients a little, as compared with Figure 11(a), and so increases I_C and the value of β, but only by a little.

Figure 11(c) shows the effect of a somewhat larger collector reverse bias. The increased reverse voltage across the collector-base junction causes its transition region to widen (as described in Section 3.2) so that the base region becomes narrower. This increases the electron density gradient in the base region a little more, and so causes a small further increase in both I_E and I_C.

Thus a graph of I_C versus V_{CB}, when V_{BE} is held constant, is as shown in Figure 11(d). The points labelled on the graph refer to the currents in diagrams (a), (b) and (c) above. The collector and emitter currents are not constant as the collector voltage is increased, but they increase gradually as the base region becomes narrower due to the increasing reverse collector voltage. This phenomenon is called the **Early effect** after J. M. Early of the Bell Laboratories who first drew attention to it.

☐   In each of Figures 11(a) to 11(c) the minority carrier density just next to the emitter-base transition region is the same. Why is this?

■   This is because the voltage V_{BE} is being kept constant. Remember that it is the voltage applied to a junction that determines minority carrier density just next to the transition region.

If the emitter current I_E were to be held constant (instead of constant V_{BE} as in Figure 11)—for example by replacing the 0.65 V source in Figure 10 with a larger voltage source in series with a resistor, as shown in Figure 12(a)—the changes in electron density profile with increasing V_{CB} are different.

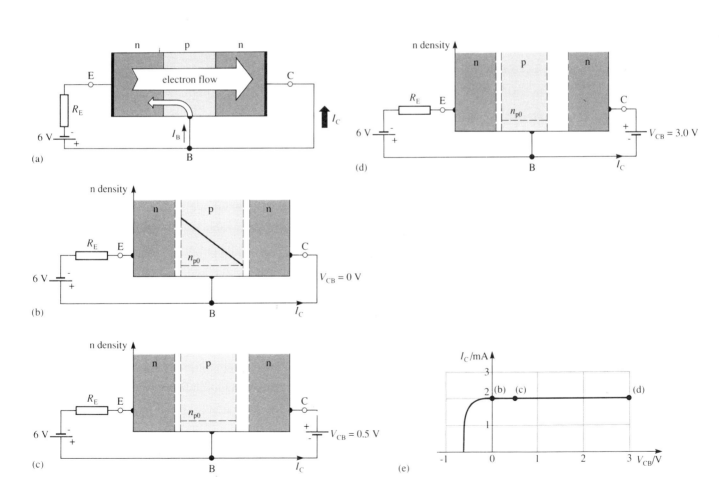

FIGURE 12 The same as Figures 10 and 11 except that the emitter current, instead of the emitter-base voltage, is held constant

SAQ 7

(a) What can you conclude about the relative magnitudes of the emitter currents in the three situations depicted in Figures 11(b) to (d)?

(b) Suppose the emitter current is held constant, as in the circuit of Figure 12(a), giving the minority carrier density profile when $V_{CB} = 0$ of Figure 12(b), what would the profiles be like when first a small voltage and then a larger voltage were applied to the collector? Sketch your answer on Figures 12(c) and (d).

The graph of I_C versus V_{CB} with I_E held constant is now as shown in Figure 12(e), with points labelled (b), (c) and (d) corresponding to the currents in diagrams (b), (c) and (d) above.

The circuits in Figures 10 and 12 are called **common-base** connections because the base terminal is common to both the input and output circuits. However it is often the case that a transistor is connected in a **common-emitter** arrangement, with the emitter common to both input and output as in the circuit of Figure 13(a). A family of d.c. output characteristics for this arrangement is shown in Figure 13(b). Each characteristic curve corresponds to a different value of input voltage V_{BE}.

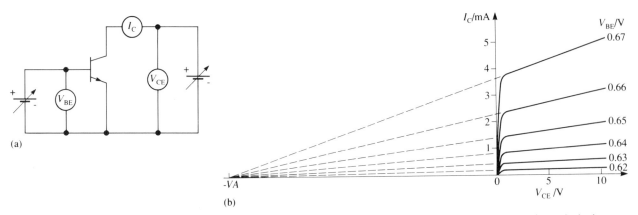

(a)

(b)

FIGURE 13 (a) A circuit for measuring the common-emitter d.c. characteristics of a transistor. The arrows through the battery symbols indicate that the applied voltages can be varied. (b) The common-emitter d.c. characteristics drawn with an exaggerated slope to show how, for $V_{CE} < 1$ V, they tend to converge on a particular voltage ($-VA$), where VA is called the Early voltage

To produce the data for such a family of curves, V_{BE} is set to a particular value, and then I_C is measured for different values of collector-emitter voltage V_{CE}. This gives the data for one characteristic curve. The measurements are then repeated for different fixed values of V_{BE}.

The resulting characteristic curves in Figure 13(b) are drawn with a rather exaggerated slope, and are extended 'backwards' until they meet the voltage axis for reasons to be explained in a moment.

☐ The curves in Figure 13(b), corresponding to equal intervals of V_{BE}, are not equally spaced. Why is this?

■ It is because I_E is an exponential function of V_{BE}. So equal steps of V_{BE} correspond to increasing steps of I_E and of I_C.

Under normal conditions of forward bias, the d.c. equation for a p–n junction (equation (8)) applies to the emitter-base junction so, neglecting the -1,

$$I_E = I_{SE} e^{KV_{BE}} \tag{12}$$

where K is again about 40 V^{-1}. So the variation of I_E with V_{BE} of a bipolar transistor is basically the normal p–n junction d.c. characteristic. However, as already explained, the Early effect does affect the emitter current somewhat (the value of I_E for a given V_{BE} is somewhat increased as V_{CE} is increased) so equation (12) applies to a specific value of V_{CE}.

Substituting values into equation (12) shows that the 10 mV increments of V_{BE} shown in Figure 13(b) correspond to increases in I_E and I_C by a factor of $\exp(40 \text{ V}^{-1} \times 0.01 \text{ V}) = e^{0.4} \approx 1.5$, as shown.

The effect of the Early effect on the transistor characteristics can be expressed in terms of the transistor's output resistance or conductance. We know that the *output conductance* of a device is, by definition, dI_{OUT}/dV_{OUT}, which is *the slope of the d.c. output characteristic*, so, to state the output conductance of the device, an expression for the slope of the d.c. characteristic is needed. As you can see from Figure 13(b), the main part of a transistor's d.c. output characteristics can be extrapolated backwards to converge on the voltage axis at a point labelled $-VA$. VA is called the **Early voltage**, and is typically between 50 V and 200 V. And from the geometry of the diagram it is evident that the slope of an output characteristic is simply $I_C/(VA + V_{CE})$.

The common-emitter output conductance is, therefore

$$\frac{I_C}{VA + V_{CE}}. \tag{13}$$

So, in summary, by specifying I_{SE} it is possible to use equation (12) to calculate I_E (at a specified value of V_{CE}) for a given value of applied V_{BE}. By specifying β it is possible to use equation (11) to calculate I_C and I_B. And by specifying the Early voltage it is possible to use equation (13) to deduce how I_C varies with V_{CE} around the calculated value of I_C for $V_{CE} > 1$ V).

(Note that the behaviour of the transistor when $V_{CE} < 1$ V is explained in Part 3 of this block.)

SAQ 8

(a) Calculate the values of I_C, I_E and I_B of a transistor in the common-emitter connection when $V_{CE} = 8$ V, given that $V_{BE} = 0.63$ V, $I_{SE} = 10^{-13}$ A at $V_{CE} = 5$ V, $\beta = 200$, $VA = 150$ V, and $K = 40$ V^{-1}.

(b) What is the output conductance of the transistor at this operating point?

SAQ 9

Referring back to Figures 11 and 12 and the effect of varying V_{CB} when first V_{BE} and then I_E is held constant, which of the two arrangements gives the higher output resistance?

The above explanations refer to n–p–n transistors; but of course the same explanations apply to p–n–p transistors except that it is holes that diffuse across the base region from emitter to collector. The polarities of the batteries must also be reversed to give the correct voltages.

Note also that the above explanations only present the essential features of bipolar transistor action. There are refinements in the construction of modern transistors, which need not concern us here, which alter the details of the explanations given. For example, in most bipolar transistors, carriers do not cross the base region solely as a result of diffusion. Consideration of the consequences of such differences is left to a later course.

Finally it should be pointed out that, as with p–n junction diodes, there is a limit to the voltages and currents that can be handled by transistors without their suffering damage. These limits, or 'ratings' are specified by manufacturers and should obviously not be exceeded.

4.3 D.C. CURRENT SOURCES

An important application of bipolar transistors is in the construction of current sources. Most electronic circuits are powered by one or two d.c. *voltage* sources, such as batteries, but sometimes, as you will see in Part 2 of this block, d.c. *current* sources are to be preferred. (You may recall that a current source was needed for measuring resistance using a digital voltmeter.) Unfortunately there are no 'natural' sources of current—as there are of voltage—so such sources have to be derived from voltage sources using specially designed circuits.

☐ What is the difference between an ideal d.c. current source and an ideal d.c. voltage source?

■ A current source is evidently the complement, or dual, of a voltage source. Just as an ideal voltage source delivers a specified voltage to any load (other than one of zero resistance) so an ideal current source delivers a specified current to almost any load.

Since a good voltage source, such as battery, has a very small internal resistance (or equivalent output resistance), it is to be expected that a good current source should have a very *large* equivalent output resistance. The following illustrative circuits are designed to supply a specified current to a load connected to a 9 V d.c. supply. They can of course be adapted for use with loads connected to other supply voltages.

A simple, but rather impractical, form of current source is shown in Figure 14(a). It consists of a large voltage source V_{SS} in series with a large resistor R_S. The resistance R_L, shown dotted, represents the resistance of the circuit that the current source is to supply with current.

In this circuit

$$I_L = \frac{9 \text{ V} - V_{SS}}{R_S + R_L}$$

and provided $R_L \ll R_S$, I_L will be virtually unaffected by changes in R_L; so for small values of R_L the circuit is essentially a constant current source. The reason why it is an impractical circuit is that it requires a very large negative voltage V_{SS} and a large resistance R_S (which may waste a lot of power) to produce a reasonably constant current. The design of practical current sources involves finding a way of providing a well specified current *without* using a very large d.c. voltage source. In other words, it involves designing d.c. circuits with large output resistances.

Figure 14(b) shows a current source which makes use of the common-base connection of a transistor. It requires two voltage sources and is therefore only suitable for some circuits, and it only allows a maximum voltage drop across R_L of about 9 V (i.e. $0 < V_{CB} < 9$ V).

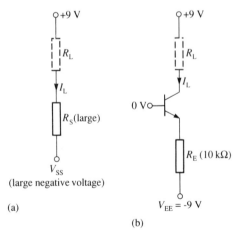

(a)

(b)

FIGURE 14 Two simple d.c. current sources: (a) a high voltage plus a high resistance; (b) using the common-base connection of a transistor

SAQ 10

(a) What current does the circuit of Figure 14(b) supply to the load resistor R_L?

(b) Remembering that output conductance is $\Delta I/\Delta V$, why does this circuit have a large output resistance?

The most commonly used current sources are the **current mirrors**, shown in Figure 15(a), and variations of it.

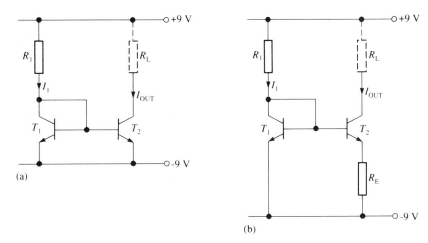

FIGURE 15 Current sources in which one transistor provides the d.c. bias for the other: (a) the current mirror in which T_2 delivers a current to the load equal to I_1; (b) the addition of a feedback resistor R_E. Here $I_{OUT} < I_1$ and the output resistance is greater than in (a)

Consider first the circuit of Figure 15(a). Its operation depends on the fact that it is relatively easy to make pairs of closely matched or nearly identical transistors on a single chip of silicon.

The collector junction of T_1 is shorted out so that $V_{CB} = 0$. But, as explained earlier, the majority of the current I_1 still flows through the collector junction, leaving a base current of only I_C/β. And because the two transistors are almost identical, the voltage drop that I_1 produces across the emitter junction of T_1 is precisely the base-emitter voltage required by T_2 to give an output current I_{OUT} equal to I_1. So the output current I_{OUT} 'mirrors' the input current I_1, and it continues to do so even when the temperature varies.

The current I_1 can be set quite accurately, despite small variations in the voltage drop across T_1, by choosing R_1 appropriately. Thus in Figure 15(a), assuming a voltage drop of 0.65 V across the emitter of T_1,

$$I_1 = \frac{18\text{ V} - 0.65\text{ V}}{R_1}$$

and this varies only a small amount if V_{BE} varies by a few millivolts. Then, to a close approximation, $I_{OUT} = I_1$. The circuit as shown will allow a voltage drop across R_L of up to about 17 V (i.e. about 1 V less than the total supply voltage) without much change in I_{out}.

☐ The output resistance of the d.c. source in Figure 15(a), though quite high, is not as high as that of the circuit of Figure 14(b). Why is this?

■ It is because in Figure 15(a) the input voltage is fixed—as in Figure 10. In Figure 14(b), however, the emitter *current* is held constant, which allows a much smaller variation of I_C as V_{CE} changes, implying a smaller output conductance (see Figure 12).

A higher output resistance can be achieved using the circuit of Figure 15(b). A small resistance R_E is put in series with the emitter of T_2 as shown, which helps to define I_E, rather than V_{BE}, thus increasing the output resistance. In addition this circuit is particularly useful when low currents are needed, because I_{OUT} in this circuit is a defined fraction of I_1 rather than the same as I_1. You can determine I_1 as a function of I_{OUT} as follows.

Suppose a current source of 10 μA is needed. If you choose R_E so that there is a voltage drop of 0.1 V across it, then R_E must be 0.1 V/10 μA = 10 kΩ.

Now since the bases of T_1 and T_2 are connected together

$$(V_{BE} \text{ of } T_1) = 0.1 \text{ V} + (V_{BE} \text{ of } T_2).$$

But, as explained in Section 3.2, a difference of 0.1 V in V_{BE} implies a ratio of $e^4 \approx 55$ between the currents of T_1 and T_2. So in this case $I_1 \approx 55 \times I_{OUT}$. Hence, if I_{OUT} is to be 10 μA with $R_E = 10$ kΩ, then $I_1 = 550$ μA.

In the circuit of Figure 15(b), with supply voltages $+9$ V and -9 V,

$$R_1 = \frac{17.35 \text{ V}}{0.55 \text{ mA}} = 31.5 \text{ k}\Omega.$$

In integrated circuits, this is a much more convenient value for R_1 than the 1.74 MΩ that would be required for the circuit of Figure 15(a). Large resistances take up too much silicon area.

SAQ 11 Suppose there is a limitation on the magnitude of the resistance that can be included in an integrated circuit version of the circuit of Figure 15(b). Design a 10 μA current source operating between d.c. supplies of $+9$ V and -9 V in which no resistor has a resistance greater than 15 kΩ.

(B) BIPOLAR TRANSISTORS
SUMMARIZED

1 A bipolar transistor is a 3-terminal device whose terminals are called emitter, base and collector. It consists of two back-to-back p–n junctions, centred on the base region, in either a p–n–p or an n–p–n configuration. In normal operation the collector junction is reverse biased and the emitter junction is forward biased. In this arrangement the collector current I_C is controlled mainly by the voltage V_{BE} applied between the base and emitter terminals.

2 A typical set of d.c. characteristics of an n–p–n transistor is shown in Figure 16. The input characteristic of Figure 16(a) is only slightly affected by variations of V_{CE}. The family of output characteristics shown in Figure 16(b) shows graphs of I_C versus V_{CE} for equal increments of V_{BE}.

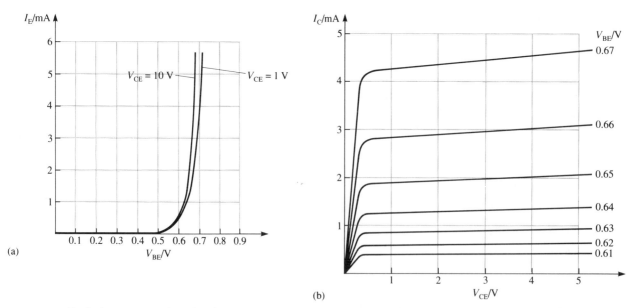

FIGURE 16 Typical common-emitter d.c. characteristics of an n–p–n bipolar transistor. (a) The input characteristic showing how it is only slightly affected by changes in V_{CE} due to the Early effect. This effect is usually ignored as a first approximation. (b) A family of output characteristics for different values of V_{BE}. Steps of 0.01 V in V_{BE} produce increases in I_C by a factor of about 1.5. The slopes of the curves as V_{CE} is increased are due to the Early effect and indicate the output conductance of the device

3 For values of V_{CE} greater than about 1 V (i.e. where they become almost straight lines in Figure 16(b)) these characteristics can be described by the two equations

(i) $$I_E = I_{SE} e^{KV_{BE}} \qquad (12)$$

where $1/K = 25$ mV. This is the forward biased p–n junction equation explained in Section 3.2 applied to the emitter-base p–n junction. The effective value of I_{SE}, the emitter saturation current, depends somewhat on the value of V_{CE}, and is typically 10^{-13} A to 10^{-15} A.

(ii) The slope of any characteristic in Figure 16(b), when V_{CE} is greater than about 1 V, is also the output conductance of the transistor and is given by the equation

$$\text{output conductance} = \frac{I_C}{VA + V_{CE}} \qquad (13)$$

where VA is the Early voltage. Typically VA lies between 50 V and 200 V.

4 Not all the emitter current I_E flows through the collector terminal; a small fraction I_B flows between emitter and base. The ratio $I_C/I_B = \beta$ is the common-emitter current gain of the transistor. The common-base current gain α is the fraction of the emitter current which flows through the collector. Evidently, since $I_E = I_C + I_B$, it follows that $\beta = \alpha/(1 - \alpha)$. Typically $\beta > 100$.

Note: (i) It follows that a particular transistor's d.c. characteristics can be specified fairly accurately, for $V_{CE} > 1$ V, in terms of just three parameters I_{SE}, β and VA.

   (ii) The behaviour of the transistor over the voltage range $0 < V_{CE} < 1$ V is discussed in Part 3 of this block.

   (iii) There are maximum limits of applied voltage and current specified by the manufacturer which must not be exceeded.

5 The high output resistance of bipolar transistors can be used as the basis for making d.c. current sources. They are usually made in the form of current mirrors, as shown in Figure 15.

6 The emitter region (of an n–p–n transistor) is more heavily doped with donors than the base region is doped with acceptors in order to give a large current gain β. Thus when the emitter–base junction is forward biased the emitter current I_E consists mainly of electrons flowing from the emitter region into the base region. A much smaller component of the emitter current consists of holes flowing from the base region into the emitter region. The ratio of the two components of I_E depends mainly on the ratio of doping densities in the emitter and base regions.

7 In normal operation the collector junction is zero or reverse biased, giving rise to a density gradient of electrons in the p-type base region causing the electrons to diffuse across it—giving rise to the collector current I_C. If recombination occurs in the base region I_C is decreased a little and I_B is increased by the same amount, with the result that β is reduced. Nevertheless β normally exceeds 100.

8 As the magnitude of the reverse bias applied to the collector junction is increased the base region narrows. If V_{BE} is held constant this causes an increase in the electron density gradient in the base region and therefore an increase in the collector and emitter currents. This is called the Early effect, and it gives rise to the transistor's output resistance (of approximately $(VA + V_{CE})/I_C$). If I_E is held constant the output resistance is higher.

5   FIELD-EFFECT TRANSISTORS

(A) DESCRIPTION AND EXPLANATION OF FET PROPERTIES

5.1 STRUCTURE

Field-effect transistors (FETs) can be regarded as three-terminal devices whose terminals are called **source**, **drain** and **gate**. There are two types of field-effect transistor, the **junction FET**, or **JFET**, and the **metal-oxide-silicon FET** or **MOSFET**.

In JFETs current flows through a channel of silicon whose cross-sectional area is controlled by the width of a p–n junction which intrudes into the channel, as illustrated in Figure 17(a). This is called the **gate**. The two ends of the channel are called **source** and **drain**. The application of a reverse bias between gate and source, causes the transition region of the gate p–n junction to widen and so reduces the width of the channel through which the current flows. In this way the applied gate-source voltage V_{GS} can be used to control the source-drain current I_D.

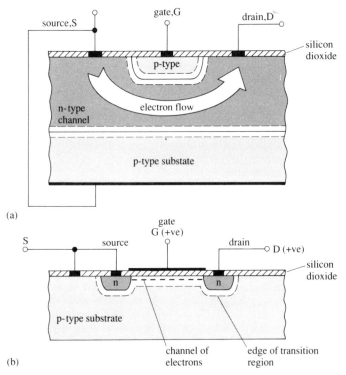

(a)

(b)

FIGURE 17 Cross-sections of field-effect transistors: (a) an n-channel JFET; (b) an n-channel MOSFET in which the channel consists of electrons induced in the transition region that is formed just under the gate

In MOSFETs the drain and source are p–n junctions formed side by side in the surface of a silicon substrate as illustrated in Figure 17(b). This time the gate is a conductor, originally a metal film (hence the name of the device), but nowadays it is usually a layer of well-doped silicon. This gate electrode is separated from the silicon substrate by a film of oxide thus forming an **input capacitance**. The application of a voltage between gate and source induces carriers in the silicon under the gate—as indicated in the diagram—the amount of charge induced in the channel being dependent on the gate voltage. When a drain-source voltage V_{DS} is applied these induced carriers flow between source and drain—the larger the induced charge the greater the drain current I_D. Thus again, but due to a quite different kind of interaction, the gate voltage V_{GS} can be used to control the source-drain current I_D. Note that in JFETs the silicon material through which the drain current flows is called the channel, but in MOSFETs it is the induced carriers, not the material in which the carriers are induced, which constitute the channel.

Both types of device can be made in either a p-channel form or an n-channel form. In **p-channel FETs** the current is carried by holes, whilst in **n-channel FETs** it is carried by electrons. Figure 17 shows n-channel devices. To begin with I shall concentrate on n-channel MOSFETs.

In an n-channel MOSFET the more *positive* the applied gate voltage the greater the density of *electrons* induced under the oxide, but the actual value of V_{GS} at which current starts to flow—called the **threshold voltage** V_T—can be adjusted during manufacture, as I shall explain. If the threshold voltage is positive (in an n-channel device) the device is called an **enhancement-mode MOSFET**. In such devices the drain current is therefore zero when $V_{GS} = 0$. If the threshold voltage is negative (in an n-channel device) the device is called a **depletion-mode MOSFET**, and $I_D \neq 0$ when $V_{GS} = 0$. I shall concentrate initially on enhancement-mode n-channel MOSFETs.

FETs are actually four-terminal devices, the substrate forms the fourth element to which a connection is made. However the substrate is often connected directly to the source, as illustrated in Figure 17, which is the form in which the devices will be discussed here.

5.2 OPERATION OF MOSFETS

Enhancement-mode n-channel MOSFETs operate as follows. The explanations of depletion-mode and p-channel MOSFETs are simple extensions of the same explanation. The operation of JFETs is briefly explained in the next section.

First note that connecting the source to the substrate as already described gives a zero voltage bias to the source-substrate p–n junction, so that no current normally flows through it. It also means that applying an input voltage V_{GS} between gate and *source* also applies a voltage between gate and *substrate* (i.e. across the capacitance formed by the gate electrode and the substrate). A positive voltage V_{GS} in excess of the threshold voltage will place positive charges on the gate electrode, and negative charges— namely electrons—will be induced in the substrate. These electrons form the channel. This is the state of affairs illustrated in Figure 18(b). The V_{GS} of 5 V exceeds the V_T of 2 V in this illustration, so a channel of electrons has been formed. A small drain-source voltage V_{DS} has also been applied so that a small drain current flows.

The main reason for there being a threshold voltage (i.e. electrons are not induced in the channel immediately a positive V_{GS} is applied) is as follows. Since the substrate is made of p-type material, the carriers present in the substrate to begin with are holes. The first volt or so of applied gate voltage is therefore concerned with repelling holes from under the oxide, thus creating a layer of silicon just under the oxide which is almost depleted of carriers. In fact a transition region like that in a p–n junction is produced. Electrons will not be induced into the substrate until the holes have been driven back and a transition region formed, as indicated in Figure 18(a). A further increase of the gate voltage drives the holes even further away, but also attracts some electrons from the source to form the 'channel' of electrons under the oxide as already described. Once the threshold voltage has been exceeded, the density of the electrons created is approximately proportional to the extra gate voltage applied.

It is because it is an *electric field* which creates the channel that the devices are called field-effect transistors. In most textbooks the channel is called an 'inversion layer', which seems to imply that the silicon has been converted to n-type by the applied electric field. But this of course is impossible; the

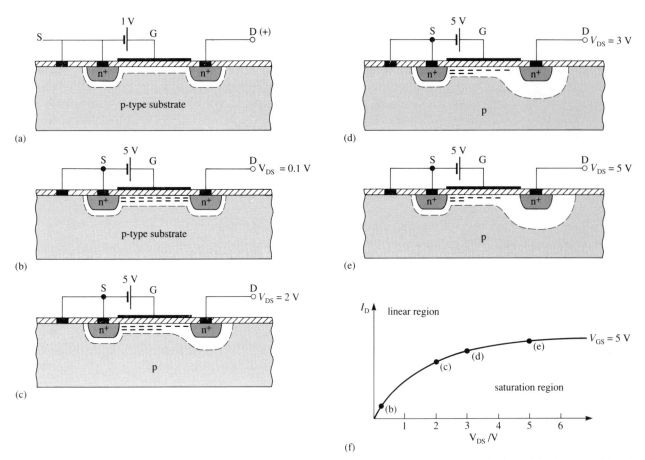

FIGURE 18 The operation of an n-channel MOSFET ($V_T = 2$ V) (a) When $V_{GS} < V_T$ (e.g. $V_{GS} = 1$ V) a 'transition' region which is almost depleted of holes is created under the gate. When V_{GS} is held at 5 V, diagrams (b), (c), (d) and (e) show cross-sections of the device as the drain voltage is increased. Diagram (d) shows the pinch-off point. Note that the transition region, shown white, widens and that the density of electrons at the drain end of the channel decreases as V_{DS} is increased. Beyond pinch-off, as in (e), the channel shortens as the transition region widens. Diagram (f) shows the d.c. characteristic that results

substrate still contains acceptors so it is p-type materal. All the term 'inversion layer' means is that the majority carriers have been removed from the surface layer and replaced by a thin layer of electrons.

Once the channel has been created, a positive voltage applied between drain and source—which is at right angles to the field due to the gate and so does not interfere with it—draws electrons from the source through the channel to the drain, thus creating the drain current I_D.

Figures 18(b), (c), (d) and (e) show what happens as the drain voltage is increased. When V_{DS} is very small, as in Figure 18(b), the voltage between the gate and the substrate is almost the same at all points in the channel, namely 5 V. Therefore, since the density of electrons at any point in the channel is a function of the voltage between gate and substrate, the density of electrons in the channel is about the same all along the channel, as indicated.

However, as V_{DS} is increased the voltage difference between gate and drain decreases, so the electron density in the channel decreases at the drain end, though it stays the same at the source end. So the density of electrons in the channel tapers down towards the drain end of the channel, as indicated in Figure 18(c). For a given gate voltage, therefore, the number of electrons in the channel decreases as the drain voltage increases, which means that although the drain current increases with increasing drain voltage it does not increase in proportion to V_{DS}; instead it tends to level off. A graph of I_D versus V_{DS} is shown in Figure 18(f). Points (b) and (c) on the graph correspond to Figures 18(b) and (c).

Figure 18(d) shows **pinch-off** occurring. Here the voltage difference between gate and drain has dropped to the threshold voltage so that at the drain end of the channel there are no induced electrons. The channel is said to be pinched-off. Pinch-off occurs when $V_{GS} - V_{DS} = V_T$. In this illustration $V_{GS} = 5$ V, $V_{DS} = 3$ V and $V_T = 2$ V. Point (d) in Figure 18(f) is the **pinch-off point**. The drain current is not however cut-off at the pinch-off point, as you might at first expect; all that happens is that the drain end of the channel becomes part of the transition region of the drain p–n junction,. Electrons in the p-region of a reverse biased p–n junction can always flow through it.

As V_{DS} is increased beyond pinch-off the channel gets slightly shorter as the drain transition region widens, as shown in Figure 18(e). But since the voltage drop along the channel remains the same (namely $V_{GS} - V_T$), the current through it increases, but only slightly. The effect is very similar to the Early effect in bipolar transistors. Point (e) in Figure 18(f) corresponds to the diagram in Figure 18(e).

Summarizing then, with V_{GS} held constant at above the threshold voltage, an increasing drain voltage causes an increasing drain current, but one that tends to flatten off because the channel contains progressively fewer electrons. This part of the characteristic, below the pinch-off point, is called the **linear region of operation**. (The reason for this name is simply that for a fixed *small* value of V_{DS} the drain current is proportional to V_{GS}.) Beyond pinch-off the amount of charge in the channel and the voltage drop along it remain almost the same, so I_D doesn't increase much as V_{DS} is increased. The extra drain voltage appears across the drain p–n junction rather than between the ends of the channel. This region is called the **saturation region of operation**. The small increase in I_D that does occur is due to a reduction of the channel length as the drain transition region widens.

Figure 19 shows a family of d.c. characteristics for different values of V_{GS}. As with bipolar transistors, if you extrapolate such a family of output characteristics backwards they tend to converge and intercept the voltage axis at about the same point. The magnitude of this voltage is referred to as $1/\lambda$ where λ is called the **channel length modulation factor**, and is typically about 0.02 V^{-1}. From the geometry of this diagram it is clear that the slope of a characteristic in the saturation region is $I_D/(V_{DS} + 1/\lambda)$. As with bipolars, this slope is the output conductance of the device, so

$$\text{output conductance (saturation region)} = \frac{I_D}{V_{DS} + 1/\lambda} \qquad (14)$$

Note that the output conductance in the linear region is considerably greater than this since the slope of the characteristic is much greater.

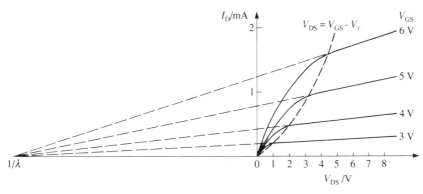

FIGURE 19 A family of n-channel MOSFET characteristics, showing their tendency to converge at a voltage of $-1/\lambda$ on the V_{DS} axis

Figure 19 also shows that the curves are not equally spaced, for reasons which will emerge when you do SAQ 13. It also shows the locus of the pinch-off points, which is given by the equation $V_{DS} = V_{GS} - V_T$.

☐ What is the threshold voltage of the device whose characteristics are plotted in Figure 19?

■ The value of V_{GS} at which the drain current just equals zero is evidently 2 V, to the nearest volt. Also, at the pinch-off points $V_{DS} = V_{GS} - V_T$. Normally the locus of pinch-off points is not marked on a family of output characteristics, but you can estimate its position by noting where the curves tend to level off.

SAQ 12 (i) Why is there no significant diffusion current between source and drain in the channel of a MOSFET?

(ii) Why is there no significant drift current in the base region between emitter and collector of the bipolar transistors described in Section 4?

It can be shown that the equation describing the d.c. output characteristic in the *linear region of operation*, where $V_{DS} \leqslant V_{GS} - V_T$, is

$$I_D = \beta V_{DS}(V_{GS} - V_T - V_{DS}/2) \tag{15}$$

where β is the **gain factor** and depends on the width and length of the channel as well as on the thickness of the oxide layer covering the channel. (This β, which is further explained in part 2 of this block, has nothing to do with the current gain of bipolar transistors.)

And in the saturation region of operation, where $V_{DS} \geqslant V_{GS} - V_T$,

$$I_D = \frac{\beta(V_{GS} - V_T)^2(1 + \lambda V_{DS})}{2} \tag{16}$$

SAQ 13 (a) On the blank graph paper of Figure 20, plot the output characteristics for a MOSFET with the following parameters: $\beta = 6 \times 10^{-4}$ A V^{-2}, $\lambda = 0.04$ V^{-1}, $V_T = -2$ V. Plot graphs for $V_{GS} = -1$ V, 0 V and +1 V.

(b) What is the output resistance of the device when $V_{DS} = 8$ V and $I_D = 1$ mA?

(c) What kind of MOSFET do these data apply to?

(*Note:* It is usual when calculating d.c. output characteristics from equations (15) and (16) to multiply equation (15) by the factor $(1 + \lambda V_{DS})$ even though the channel length is not being modulated in this region. This is done in order to avoid discontinuities between the two equations. Its effect is small in the linear region where V_{DS} is small. You should do this too.)

As you should have deduced when doing the above SAQ, the device whose characteristics you have plotted in Figure 20 is a depletion-mode MOSFET. But how does it come about, in view of the explanation given earlier of the cause of the threshold voltage, that an n-channel device can have a negative threshold voltage?

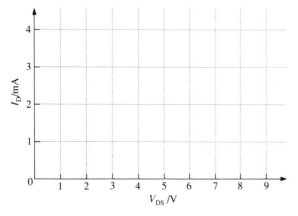

FIGURE 20 See SAQ 13

The threshold voltage can be reduced, and even made negative, by implanting a very thin layer of donors in the surface of the p-type substrate just under the gate, as indicated in Figure 21. This is done simply by firing donor atoms in a vacuum at the silicon surface. If the implanted donor density exceeds the density of holes already there, a channel of electrons will be formed even when $V_{GS} = 0$ and a drain current will flow as soon as V_{DS} is applied. Indeed a negative voltage (e.g. $V_{GS} < -3$ V) has to be applied to the gate to drive these electrons away and reduce I_D to zero. Thus V_T is negative and the device becomes a depletion-mode MOSFET.

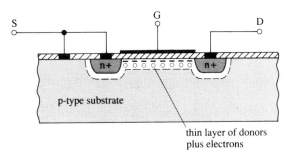

FIGURE 21 A cross-sectional diagram of an n-channel depletion-mode MOSFET showing the layer of donors implanted in the surface of the p-type substrate to form a channel of electrons even when $V_{GS} = 0$

The drain current that flows in a depletion-mode MOSFET when $V_{GS} = 0$ is called I_{DSS} and is quoted in data sheets.

☐   Why is I_{DSS} not quoted in data sheets for enhancement-mode MOSFETs?

■   Because $I_D = 0$ when $V_{GS} = 0$ in enhancement-mode MOSFETs!

The explanation of the operation of p-channel MOSFETs is identical to that of the n-channel MOSFET just described except that all region types, carrier types, voltages and currents are reversed. For example, the gate is made increasingly *negative* to cause an increasing drain current, and the threshold voltage of an enhancement-mode p-channel device is negative. Idealised families of d.c. characteristics of all types of FET are shown in Figure 25 which accompanies the summary at the end of the section.

The main electrical advantage of MOSFETs over bipolar transistors is that their d.c. gate current is virtually zero. This of course is due to the presence of the oxide layer between the gate electrode and the substrate. The current gain, if it were ever referred to, would be approaching infinity. This means that the input power to the device can be very small indeed. The advantage of MOSFETs from a production point of view is that in general they are smaller and cheaper to manufacture than bipolars. Their main disadvantage is that their transconductance g_m is normally much less than that of bipolars at the same operating current; that is the control of output current by the input voltage is normally less effective. It is possible, however, using special construction methods, to produce MOSFETs whose g_m is comparable to, or even greater than, that of bipolars.

5.3 OPERATION OF JFETS

The effect of increasing the drain voltage of an n-channel JFET is shown in Figure 22. This shows cross-sections of an n-channel device at four values of drain voltage, whilst V_{GS} is held at −5 V. Notice immediately that the gate voltage, unlike n-channel MOSFETs, is always zero or negative. This ensures that the gate p–n junction is never forward biased. The gate is made less negative, thus widening the channel, to increase the drain current. The negative voltage which completely closes the channel and prevents current flowing is the **pinch-off voltage** V_P. It is analogous to the threshold voltage of MOSFETs.

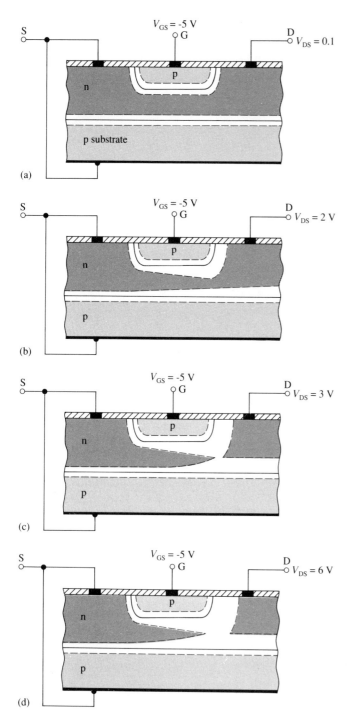

FIGURE 22 Cross-sectionl diagrams of an n-channel JFET as V_{GS} is held at -5 V and V_{DS} is increased. The gate transition region widens right through the channel at pinch-off, as shown in (c). Beyond pinch-off the channel shortens, as in (d). Note that the transition region of p–n junction between substrate and channel also widens

In Figure 22(a) $V_{DS} = 0.1$ V, and the transition region of the gate p–n junction has almost the same width everywhere because the voltage drop across it is about 5 V at all points. At $V_{DS} = 2$ V however (Figure 22(b)) the voltage drop across the gate junction is 7 V at the drain end, but still only 5 V at the source end, so the channel through which the drain current flows has become narrower at the drain end. At $V_{DS} = 3$ V (Figure 2(c)) the gate has widened right through the channel at the drain end, 'pinching it off'. Evidently in this example the pinch-off voltage $V_P = 8$ V.

Figure 22(d) shows the situation in the region beyond pinch-off, which is again called the saturation region of operation, in which the current increases more slowly than in the linear region. The reason for this slight increase is the same as in the case of MOSFETs: the channel gets shorter as the gate transition region at the drain end becomes wider due to the increased reverse voltage across it.

The d.c. characteristics of a JFET are very like those of a MOSFET even though its mode of operation is very different. Indeed, to a first approximation, the same curves of Figure 20 apply to n-channel JFETs too except that the values of V_{GS} are different, as shown in Figure 23. The output conductance can again be expressed as $I_D/(V_{DS} + 1/\lambda)$.

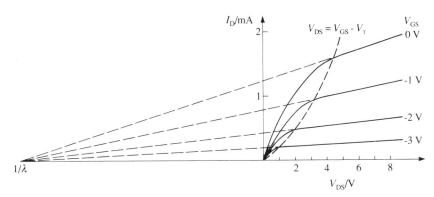

FIGURE 23 A family of n-channel JFET characteristics, showing again their tendency to converge at a voltage of $-1/\lambda$ on the V_{DS} axis

The actual equations describing the d.c. characteristics of a JFET are much more complicated than those for MOSFETs and need not concern us here.

5.4 D.C. CURRENT SOURCES USING FETS

The simplest type of current source that uses a depletion-mode MOSFET is the circuit shown in Figure 24(a). The gate is simply connected to the source, as shown. The current that flows, provided V_{DS} is sufficient to cause saturation, is I_{DSS}.

☐ Why use a depletion-mode MOSFET rather than an enhancement-mode one?

■ A depletion-mode MOSFET passes a current I_{DSS} when $V_{GS} = 0$, so it is only necessary to make $V_{GS} = 0$, as shown, to establish a current I_L. Using an enhancment mode MOSFET you would have to connect a positive voltage to the gate in order to establish a drain current.

The value of I_{DSS} is not very accurately controlled in manufacture so devices have to be selected to give the required current. A JFET can alternatively be used in this circuit; indeed 'current regulator diodes' can be bought which are simply JFETs with their gate and source connected together. The devices are selected to give different specified values of I_{DSS}.

The output resistance of the circuit shown in Figure 24(a) is simply the output resistance of the transistor, namely the reciprocal of the slope of the d.c. characteristic at the operating point, as specified by the parameter λ. Indeed you can regard the voltage $(-1/\lambda)$ as the voltage you would need to create the same current source simply by using a voltage source and a resistor as in Figure 14(a).

An improvement on the circuit of Figure 24(a), as far as the control of I_L is concerned, is shown in Figure 24(b). Here a resistor is included in the source lead through which I_L flows, creating a voltage drop which, in effect, creates a negative value of V_{GS}. The presence of R_S reduces the current through the transistor but tends to stabilize it: the larger the value of I_{DSS} the greater the voltage drop across R_S and the greater the negative value of V_{GS}. In other words the greater the value of I_{DSS} the more it is reduced! This is a form of negative feedback in which the feedback is

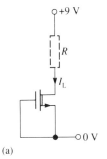

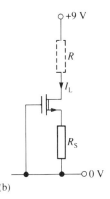

FIGURE 24 MOSFET d.c. current sources using a depletion-mode MOSFET: (a) without feedback; (b) with feedback resistor R_S included

'current derived'; the presence of R_S therefore increases the output resistance of the circuit.

 SAQ 14 The depletion-mode transistor whose characteristics you have drawn in Figure 20 is used in the circuit of Figure 24(b). The circuit is required to supply a current of about 0.4 mA to a load when the drain voltage of the transistor might be anywhere between about 2 V and 7 V. What resistance should R_S have to achieve this result?

Current mirrors using MOSFETs, similar to those using bipolar transistors (Figure 15), can also be made provided the threshold voltages of the two MOSFETs are well matched. This, however, is not so easy to achieve with MOSFETs or JFETs as it is with bipolars, since it involves careful control of a number of highly variable factors.

(B) FIELD-EFFECT TRANSISTORS SUMMARIZED

1 Field-effect transistors (FETs) can be regarded as three-terminal devices whose terminals are called source, drain and gate. There are two types of field-effect transistor, the junction FET, or JFET, and the metal-oxide-silicon FET or MOSFET. In a JFET current flows through a channel of silicon whose cross-section area is controlled by a p–n junction whose width is varied by the application of a voltage between gate and source, as illustrated in Figure 17(a). In a MOSFET the drain and source are p–n junctions formed side by side in the surface of a silicon substrate as illustrated in Figure 17(b). The gate is separated from the silicon substrate by a film of oxide. The application of a voltage between gate and source induces carriers in the silicon under the gate which then forms the channel between source and drain.

2 There are complementary forms of both types of FET: namely n-channel and p-channel devices. In n-channel devices the drain current is carried by electrons, whilst in p-channel devices it is carried by holes. There are two variants of MOSFET called depletion-mode devices and enhancement-mode devices. In enhancement-mode devices $I_D = 0$ when $V_{GS} = 0$. In depletion mode devices (and in JFETs) $I_D = I_{DSS}$ when $V_{GS} = 0$.

3 The families of output characteristics of all FETs have the same general form and are summarized in Figure 25. Note particularly that the polarities of the voltage and directions of currents of p-channel devices are the reverse of those of n-channel ones; and that the differences between the three types of n-channel devices, or between the three p-channel devices, is in their range of values of V_{GS}. In particular the threshold voltages V_T distinguish between them. (The threshold voltages are the gate voltages at which the drain current I_D just begins to flow.)

4 Each set of characteristic curves can be divided into the 'linear' region and the 'saturation' region of operation. In the linear region the characteristics are very curved (!) whilst in the saturation region they are almost straight. The slope of the output characteristic is equal to the output conductance of the transistor. In the saturation region, this is given by

$$\text{output conductance} = \frac{I_D}{V_{DS} + 1/\lambda} \tag{14}$$

where λ is called the channel length modulation factor. Note that the output resistance (i.e. the reciprocal of the output conductance) is much larger in the saturation region than in the linear region.

5 There are alternative graphical symbols for MOSFETs in common use, which are also shown in Figure 25. Note that if the (longer) central line represents the piece of silicon, an arrow in a diagram always points at a p-region or away from an n-region, as with bipolar transistors. However the arrow indicates the *source* in a MOSFET, but indicates the *gate* in a JFET.

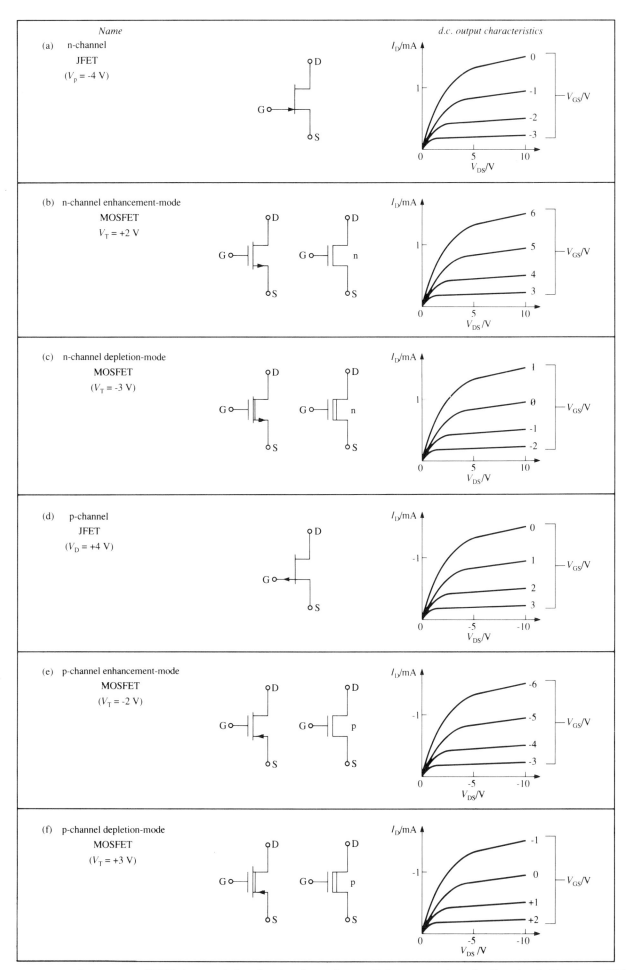

FIGURE 25 A summary of FET characteristics, showing, for each type, (i) its name and pinch-off or threshold voltage, (ii) its graphical symbol or icon, and (iii) the form of a typical set of d.c. characteristics

Note the extra line added to the MOSFET symbols for depletion-mode operation—it is intended to indicate the existence of a channel when $V_{GS} = 0$. Other symbols may be found in data sheets and in textbooks, so be sure to check the meanings of symbols in other publications. In particular a more complicated standard symbol is used to represent the four-terminal nature of MOSFETs, but we shall not be referring to it in this text.

6   The carriers in the channel of either type of FET flow from source to drain under the influence of an electric field, so the currents are drift currents rather than diffusion currents.

7   The d.c. equations for the output characteristics of a MOSFET are given as equations (15) and (16). For the JFET the equations are more complicated, but describe characteristics that are similar in form.

8   FET current sources can be made by connecting the gate to the source of a JFET or a depletion-mode MOSFET. The current that flows is I_{DSS}. The performance can be improved by the use of current-derived feedback.

ANSWERS TO SELF-ASSESSMENT QUESTIONS

SAQ 1

From equation (1), resistance = length/(conductivity × cross-sectional area). Therefore

(i) $R = \dfrac{100 \text{ m}}{(0.00025)^2 \text{ m}^2 \times \pi \times 5.8 \times 10^7 \text{ S m}^{-1}} = 8.78 \ \Omega$;

(ii) $100 \ \Omega = \dfrac{5 \times 10^{-3} \text{ m}}{\sigma \times 5 \times 10^{-4} \text{ m} \times 2 \times 10^{-3} \text{ m}}$

so $\sigma = 50 \text{ S m}^{-1}$.

SAQ 2

(a) (i) The donors and acceptors cancel one another out on an atom by atom basis. Each electron released into the crystal by a donor atom is mopped up by an acceptor atom. The silicon then behaves almost as if it had not been doped at all. It is said to be *compensated*. (ii) The type of dopant that is in the majority determines whether the silicon is n-type or p-type. If donors dominate the material is n-type, for example. The silicon behaves almost as if it had been doped only with the one dopant to a density equal to the difference between the densities of the two types of dopant.

(b) Phosphorus is a donor so $N_d = 10^{21} \text{ m}^{-3}$. Boron is an acceptor so $N_a = 5 \times 10^{20} \text{ m}^{-3}$. The net doping density is $N_d - N_a = 10^{21} - 5 \times 10^{20} = 5 \times 10^{20} \text{ m}^{-3}$, since the density of phosphorus is greater than that of boron. So conductivity $= 2.4 \times 10^{-20} \times 5 \times 10^{20} = 12 \text{ S m}^{-1}$.

SAQ 3

(i) In intrinsic material both holes and electrons contribute to the conduction process, so $\sigma = qn_i(\mu_n + \mu_p) = 1.6 \times 10^{-19} \times 1.5 \times 10^{16} \times (0.15 + 0.045) = 4.7 \times 10^{-4} \text{ S m}^{-1}$.
(ii) The conductivity of silicon increases by the factor 1.08 for every degree centigrade rise of temperature. Thus if the temperature increases from 25°C to 100°C, the conductivity increases by the factor $(1.08)^{75} = 321$. The conductivity at 25°C is $4.7 \times 10^{-4} \text{ S m}^{-3}$, so the conductivity at 100°C is $4.7 \times 10^{-4} \times 321 = 0.15 \text{ S m}^{-1}$.

SAQ 4

The equilibrium density of holes in the n-region is given by equation (5). Thus $p_0 = n_i^2/N_d = (1.5 \times 10^{16})^2 \div 10^{22} = 2.25 \times 10^{10} \text{ m}^{-3}$. The factor by which this is increased by the forward bias is $\exp(KV_D) = \exp(40 \times 0.6) = e^{24} = 2.65 \times 10^{10}$. Therefore the hole density next to the transition region is

$$2.25 \times 10^{10} \text{ m}^{-3} \times 2.65 \times 10^{10} = 5.96 \times 10^{20} \text{ m}^{-3}.$$

SAQ 5

Substituting the given values for I_S and V_D in $I_D = I_S(e^{KV_D} - 1)$ when $1/K = 25 \text{ mV}$ gives: (i) 14.5 mA; (ii) 1.96 mA; (iii) 0.26 mA; (iv) 0.089 μA; (v) $-0.98 \times 10^{-8} \ \mu$A; (vi) $-10^{-8} \ \mu$A.

In practice under reverse bias the measured current is usually greater than the above figures because of leakage currents across the surface of the junction mounting, etc.

SAQ 6

Substituting the given data into equation (8) gives $1 \text{ mA} = 10^{-13} \text{ A} \times \{\exp(35V_D) - 1\}$. The (-1) in the bracket is negligible compared with the exponential, so it can be ignored. Therefore

$$10^{-3}/10^{-13} = 10^{10} = \exp(35V_D)$$
$$35V_D = \log_e 10^{10} = 23.02.$$

So

$$V_D = 23.02/35 = 0.658 \text{ V}.$$

If the forward voltage is 0.4 V, equation (8) becomes

$$I_D = 10^{-13}\{\exp(35 \times 0.4) - 1\}$$
$$= 10^{-13} \times e^{14}$$
$$= 10^{-13} \times 1.2 \times 10^6$$
$$= 1.2 \times 10^{-7} \text{ A} = 0.12 \ \mu\text{A}.$$

SAQ 7

(a) Since the gradient of the minority carrier density in the base region increases as the reverse bias on the collector junction is increased, the emitter electron current, and therefore I_E as whole, must increase too.

(b) The minority carrier density gradient must remain the same in the base region since I_E does not change. The required electron density profiles, as V_{CB} increases, must therefore be as shown in Figure 26(b) and (c). The *gradients*

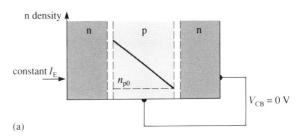

(a)

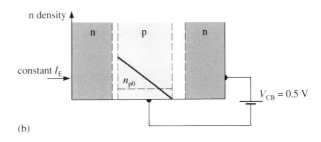

(b)

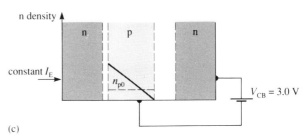

(c)

FIGURE 26 The electron density profiles in the base region of an n–p–n transistor as V_{CB} is increased and I_E is held constant. See answer to SAQ 7

are the same as in Figure 12(b), which is reproduced in Figure 26(a). Note that since the electron *density* next to the emitter junction decreases as V_{CB} increases, the forward bias needed to be applied to the emitter decreases a little.

SAQ 8

(a) First calculate I_E from equation (12). Thus $I_E = 10^{-13}$ A $\times e^{40 \times 0.63} \approx 8.8 \times 10^{-3}$ A $= 8.8$ mA at $V_{CE} = 5$ V.

Since $\beta = 200$, then $\alpha = 0.995$, so, by equation (11), $I_C = 8.76$ mA and $I_B = 0.044$ mA at $V_{CE} = 5$ V.

Finally, by equation (13), when $VA = 150$ V

$$\frac{8.76 \text{ mA}}{150 \text{ V} + 5 \text{ V}} = \frac{I_C \text{ when } V_{CE} = 8 \text{ V}}{150 \text{ V} + 8 \text{ V}} \text{ for constant } V_{BE}.$$

So

$I_C = 8.9$ mA at $V_{CE} = 8$ V.

(b) The output conductance is given by equation (13) again, so

$$\text{output conductance} = \frac{8.9 \text{ mA}}{158 \text{ V}} = 56.3 \ \mu\text{S}.$$

SAQ 9

If you compare the electron density gradients in Figures 11 and 26 (the answer to SAQ 7) you will see that whereas the gradient changes significantly when V_{BE} is held constant, it changes very little when I_E is held constant. Thus when I_E is held constant the output current changes less so the output resistance is greater.

SAQ 10

(a) The output current is almost the same as the emitter current ($I_C = \alpha I_E$), and the emitter current is evidently $(-0.65 \text{ V} - V_{EE})/R_E = 8.35 \text{ V}/10 \text{ k}\Omega = 0.835$ mA.

(b) The output resistance of the circuit is high because, as explained in SAQ 9 changes in V_{CB} have very little effect on I_C when I_E is held constant even though the base width changes.

SAQ 11

You can conclude from the worked example that the larger you make R_E, the smaller you can make R_1. Suppose therefore we make $R_E = 12$ kΩ, so that the voltage drop across it is to be 0.12 V. This implies that $I_1 = I_{OUT} \times \exp(0.12 \text{ V}/0.025 \text{ V}) = I_{OUT} \times e^{4.8}$.

But $e^{4.8} = 121.5$, so $I_1 = 1.215$ mA.

Therefore, $R_1 = 17.35 \text{ V} \div 1.215 \text{ mA} = 14.3$ kΩ. Both resistors are less than 15 kΩ as required. Any value of R_E between about 12 kΩ and 15 kΩ will satisfy the specified requirement.

SAQ 12

(i) There is virtually no diffusion current in a MOSFET because there is no density gradient of minority carriers to speak of. As explained in the next section a small gradient is developed, but it has negligible effect.

(ii) The reason why there is little drift current in the base region of the bipolar transistors described in Section 4 is because, unlike the channel in a MOSFET, the base region in a bipolar is full of majority carriers, so it is not possible to develop much of an electric field in it. The very large density

gradient of minority carriers dominates the behaviour of such devices.

(It is worth noting that modern bipolar transistors are constructed in such a way that the flow of carriers through the base region is enhanced by the presence of an electric field, but how this is done need not concern us in this course.)

SAQ 13

(a) For a transistor in which $\beta = 6 \times 10^{-4}$ A V^{-2}, $\lambda = 0.04$ V^{-1}, $V_T = -2$ V the following data can be calculated from equations (15) and (16).

Linear region: $I_D = \beta V_{DS}(V_{GS} - V_T - V_{DS}/2)(1 + \lambda V_{DS})$ where $V_{DS} \leqslant V_{GS} - V_T$.

V_{DS}/V	V_{GS}/V	$1 + \lambda V_{DS}$	V_{DS} $(V_{GS} - V_T$ $- V_{DS}/2)$	I_D/mA
0.5	−1	1.02	0.5 × 0.75	0.23
1.0	−1	1.04	1.0 × 0.5	0.31
1.0	0	1.04	1.0 × 1.5	0.94
1.5	0	1.06	1.5 × 1.25	1.19
2.0	0	1.08	2.0 × 1.0	1.30
1.0	+1	1.04	1.0 × 2.5	1.56
2.0	+1	1.08	2.0 × 2.0	2.59
3.0	+1	1.12	3.0 × 1.5	3.02

Saturation region: $I_D = (\beta/2) \times (V_{GS} - V_T)^2(1 + \lambda V_{DS})$ where $V_{DS} \geqslant V_{GS} - V_T$.

V_{DS}/V	V_{GS}/V	$1 + \lambda V_{DS}$	$(V_{GS} - V_T)^2$	I_D/mA
1.0	−1	1.04	1	0.31
8.0	−1	1.32	1	0.40
2.0	0	1.08	4	1.30
5.0	0	1.2	4	1.44
8.0	0	1.32	4	1.58
3.0	+1	1.12	9	3.02
5.0	+1	1.2	9	3.24
8.0	+1	1.32	9	3.56

The characteristics of Figure 27 are a plot of these data.

(b) You can find $1/\lambda$ either by extrapolating the graphs of Figure 27 back to the voltage axis as in Figure 19, or else by using equation (14) as follows. Consider the graph for $V_{GS} = 1$ V. Substituting the data for $V_{DS} = 8$ V and $V_{DS} = 3$ V in equation (14) gives

$$\frac{3.56}{1/\lambda + 8 \text{ V}} = \frac{3.02}{1/\lambda + 3 \text{ V}},$$

so $(3.56 - 3.02) \times 1/\lambda = 8 \times 3.02 - 3 \times 3.56$

Therefore $1/\lambda = 13.48/0.54 \approx 25$ V.

Using equation (14) again with this calculated value of $1/\lambda$ gives the required output resistance r_0:

$$r_0 = \frac{25 \text{ V} + 8 \text{ V}}{1 \text{ mA}} = 33 \text{ k}\Omega.$$

(c) The transistor is evidently an n-channel device (because V_{DS} is positive), and it is a depletion-mode device because $I_D \neq 0$ when $V_{GS} = 0$.

SAQ 14

If the transistor is to supply about 0.4 mA the transistor must operate according to the lower of the three curves you have drawn in Figure 20; see also Figure 27. That is, V_{GS} should be -1 V. If R_s is set at 2.5 kΩ there will be 1 V drop across it when 0.4 mA flows. This will, in effect, apply the required value of V_{GS} and so the required current will flow.

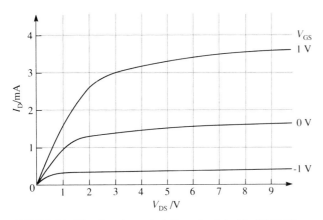

FIGURE 27 The d.c. characteristics calculated in SAQ 13

INDEX OF KEY TERMS

PART 2   ANALOGUE TRANSISTOR CIRCUITS

CONTENTS

AIMS

1 To explain how the equivalent circuits of bipolar transistors and MOS-FETs are related to transistor structure.

2 To explain in principle how various basic transistor circuits operate so that their performance can be understood.

3 To show how a simple computer circuit simulation program can be used to calculate the performance of transistor circuits.

OBJECTIVES

SPECIFIC OBJECTIVES:

After completing this text you should be able to:

1 Relate the low-frequency equivalent-circuit parameters of a transistor at a particular operating point to specified transistor data.

2 Relate the low-frequency equivalent-circuit parameters at one operating point to those at another operating point.

3 Estimate the voltage gain and the input and output resistance of a common-emitter amplifier, of an emitter-follower, of a long-tailed pair, of a CMOS amplifier and of an operational amplifier under given operating conditions.

4 Calculate the gain, the input resistance and the output resistance of a transistor circuit using a circuit simulation computer program.

GENERAL OBJECTIVES

After studying this text you should be able to:

1 Explain, in terms of transistor properties, the operation of the following basic circuits:

common-emitter or common-source amplifier
current mirror
dynamic load
emitter-follower or source-follower
long-tailed pair
a simple operational amplifier
a CMOS amplifier

2 Explain the meanings of the following new terms:
common-mode rejection ratio (CMRR)
direct-coupled circuit
push-pull output
rail rejection
short-circuit or open-circuit input or output
spread (of parameter values)
temperature compensation

1 INTRODUCTION AND STUDY GUIDE

This part of Block 4, and Part 3 as well, have a good deal of ground to cover. They are concerned with the business of making use of what you have learnt about signals and about transistors and other devices for the purpose of constructing useful transistor circuits. This text concentrates on bipolar transistors because they are more widely used in analogue circuits. MOSFETs come into their own in digital circuits, and are therefore discussed in more detail in Part 3. Figure 1 is a conceptual diagram of the coverage given in this text to bipolars. A similar, but much abbreviated plan applies to MOSFETs.

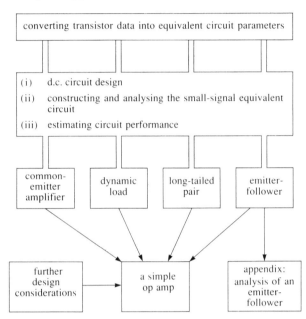

FIGURE 1 A flow chart illustrating the structure of this part of Block 4. Beginning with transistor data, four basic circuits are considered. These are then brought together in a simple op amp

The design process begins with getting to know the important details of the components you intend to use. By now you know enough about resistors, capacitors, sinusoidal waveforms, etc., but it is still necessary to translate the explanations of how transistors work into practical parameters to be used in circuit design. This is the theme of Section 2.

This information enables you to do three things: (i) it enables you to design the d.c. circuit—which means you can choose the right resistors to connect between the transistor(s) and the d.c. supplies, so that the transistors have the required quiescent terminal d.c. voltages and currents; (ii) it enables you to establish the values of the small-signal equivalent-circuit parameters of the transistors used in the circuit in preparation for a detailed nodal analysis. With simple circuits you can use algebra to do the nodal analysis, but for any circuit you can use your computer to carry out the analysis to obtain the circuit's performance; (iii) provided you understand how the circuit works, the transistor data enable you to make approximate and useful estimates of circuit performance using back-of-an-envelope calculations. In other words, if you know what you are doing you can bypass much of the nodal analysis and arrive at quite accurate estimates of circuit performance. The text explains all three approaches for each circuit discussed.

These three approaches to the design and analysis are applied to three basic circuits: the common-emitter amplifier, the emitter-follower, and the long-tailed pair. They are also applied to the current mirror which finds various uses in improved versions of the circuits described. Obviously in this introductory course you will not be expected to be able to design novel circuits, but you will be expected to be able to choose component and parameter values to give a specified performance with any one of these basic circuits. The performance parameters considered are mainly their voltage gain and input and output resistances.

As indicated in Figure 1, these circuits are brought together to form a simple op amp whose performance you should by then be able to understand fully. A few additional considerations, such as temperature compensation, become important when the high gain of an op amp is available, so these are also discussed.

The manufacturer's data are usually sufficient, as they stand, for the d.c. design of circuits, but some additional analysis sometimes needs to be done to determine small-signal equivalent-circuit parameter values, so this is the first topic discussed in the text. Manufacturers often give far more information than is needed for our purposes so I have been very selective.

2 THE SMALL-SIGNAL EQUIVALENT CIRCUITS OF TRANSISTORS

SAQ 1 *Revision* (a) What is the difference between the small-signal conductance and the d.c. conductance of a nonlinear resistor at a particular d.c. operating point? Is either conductance independent of the d.c. operating point?

(b) Figure 2 shows the d.c. characteristic of a nonlinear resistor. What are its d.c. and small-signal conductances at the d.c. operating point of 6 V?

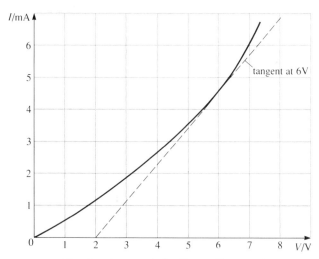

FIGURE 2   The d.c. characteristic of a nonlinear resistor

The transistor small-signal equivalent circuit that we will be using in this text is shown in Figure 3(a). It is a somewhat simplified version of the more accurate one introduced in Block 8, but it is accurate enough for now. It consists of an input conductance g_i, an output conductance g_o and a voltage-dependent current generator $g_m V_{in}$. We will be using the same form of equivalent circuit for both bipolar transistors and MOSFETs, though, of course, the values of the three parameters in the circuit are different for the two types of transistor. Sometimes it is more convenient to refer to input and output *resistances*, r_i and r_o rather than *conductances*, so either will be used in this text. Obviously $r_i = 1/g_i$ and $r_o = 1/g_o$.

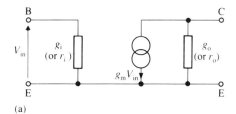

(a)

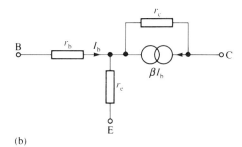

(b)

FIGURE 3 Basic, low-frequency, small-signal equivalent circuits of a transistor. (a) The 'π type' equivalent circuit which applies to both bipolar transistors and MOSFETs. It will be developed in Block 8 into the *hybrid-π equivalent circuit* for bipolar transistors. (b) The 'T-type' equivalent circuit that is also sometimes used for bipolar transistors

Any set of equivalent-circuit values applies only to one specified d.c. operating point, so it is always necessary to relate them to particular d.c. terminal voltages and currents. Using the theory developed in Part 1 of this block the parameter values at any other d.c. operating point can be estimated quite accurately, as I shall explain.

Figure 3(b) shows another equivalent circuit which is in quite common use. It is called the **T equivalent circuit** because of its T-like shape. It contains a *current*-dependent current generator, labelled βI_b, instead of the *voltage*-dependent current generator of Figure 3(a). The main disadvantages of the T circuit are that its current generator turns out to be very frequency dependent and its parameter values are not so easy to calculate, so it is not used here.

The values of the two resistors and of the transconductance g_m in the equivalent circuit of Figure 3(a) can be derived either from the appropriate d.c. characteristics of the transistor or from certain performance parameters as explained in the next two sections.

2.1 BIPOLAR TRANSISTORS

2.1.1 EQUIVALENT-CIRCUIT PARAMETERS DERIVED FROM GRAPHS

Some illustrative examples of the 'transistor characteristic curves' that manufacturers give are shown in Figure 4. The small-signal equivalent-circuit parameters are derived from the *slopes* of the appropriate characteristic curves.

(i) *Input conductance* g_i. This is the small change of base current I_B per unit small change of base-emitter voltage. That is $g_i = dI_B/dV_{BE}$, so g_i is the slope of the graph in Figure 4(a) at the chosen operating point. For example at $I_B = 20\ \mu A$ the tangent to the graph has a slope of about $80\ \mu A/0.1\ V = 0.8\ mS$, so $g_i = 0.8\ mS$ or $r_i = 1250\ \Omega$.

(ii) *Transconductance* g_m. This is the small change of the collector current I_C per unit change of V_{BE} at the chosen operating point, so $g_m = dI_C/dV_{BE}$. g_m is therefore equal to the slope of Figure 4(b) at the chosen operating point. g_m is a *trans*conductance, so the current and voltage do not refer to the same two nodes in the circuit—the current I_C flows between *collector* and emitter whilst the voltage V_{BE} is applied between *base* and emitter. So it is usually best to express g_m in units of milliamps-per-volt rather than millisiemens. Thus in this case, if $I_C = 2\ mA$, for example, the slope is $8\ mA/0.1\ V$, so $g_m \approx 80\ mA/V$ (rather than $0.08\ S$).

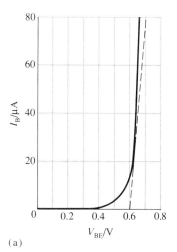

(a)

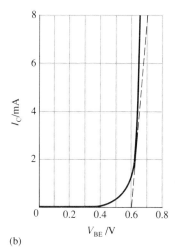

(b)

FIGURE 4 The d.c. characteristics of a bipolar transistor: (a) the input characteristic; (b) the forward transfer characteristic

Note that Figures 4(a) and 4(b) show almost the same exponential curve, but that they are plotted with different current axes. The ratio of the scales of these axes is evidently the d.c. current gain $\beta = I_C/I_B$ (in this case a factor of 100). You might therefore expect g_i and g_m to be in the ratio of β too. In fact, since g_i and g_m are *small-signal* quantities they refer to the *slopes* of these two graphs and so their ratio is not quite the same as the d.c. current gain β; their ratio is the *small-signal* current gain normally symbolised by h_{fe}. That is, dI_C/dI_B is called h_{fe}, whilst I_C/I_B is called β. However because $h_{fe} \approx \beta$ *I shall usually use β to represent both the d.c. and the low-frequency small-signal current gain of a bipolar transistor.* At higher frequencies when $h_{fe} \neq \beta$ it is of course necessary to use h_{fe} when referring to the small-signal current gain (see Block 8). Thus, although, strictly speaking $g_m = h_{fe} \times g_i$, I shall be using the approximation $g_m = \beta \times g_i$. This avoids having to refer to 'h_{fe} at low frequencies' every time I mention small-signal current gain.

☐ In view of the similarity in shape of the curves of Figures 4(a) and (b) what shape would you expect the graph of I_B versus I_C to have?

■ Since I_B and I_C are almost proportional to each other, the graph should be almost a straight line.

(iii) *Output conductance g_o.* The output conductance is the slope of the output characteristic of the transistor at the chosen operating point. Figures 4(c) and 4(d) are alternative plots of the output characteristics: In Figure 4(c) I_B is kept constant for each curve, whilst in Figure 4(d) V_{BE} is held constant for each curve.

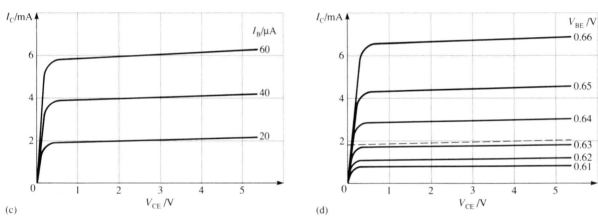

(c)

(d)

FIGURE 4 (c) the output characteristics for a family of constant values of I_B; (d) the output characteristics for a family of constant values of V_{BE}

☐ Can you see how to decide which output characteristic you should use in order to find g_o: Figure 4(c) or Figure 4(d)? (Note: You have to refer to the equivalent circuit to decide.)

■ The output conductance g_o is the conductance that you see looking into the output terminals of the equivalent circuit *when there is no current coming from the current generator*. In the equivalent circuit of Figure 3(a) the current produced by the current generator is proportional to the input signal *voltage*, so you need zero input small-signal voltage (i.e. a constant d.c. input voltage), to ensure that the current generator is not active, when determining the output conductance. So you should use Figure 4(d) in which V_{BE} is held constant.*

* *Note that if you had been using the equivalent circuit of Figure 3(b) you would have had to take the slope of the graph of Figure 4(c) in order to find r_c, because you would have to ensure that there is zero input signal current (i.e. constant d.c. base current). The slopes of the two graphs at the same operating point are not quite the same; a fact which can be represented in the equivalent circuit by the addition of a further large resistance. But this is neglected in the Figures 3(a) and (b).*

So g_o is the slope of the graph in Figure 4(d) at the chosen operating point. The dashed line shows the slope of a characteristic that would pass through the operating point of $I_C = 2$ mA and $V_{CE} = 4$ V. The slope is about 0.1 mA/5 V, giving an output conductance $g_o \approx 0.02$ mS, or an output resistance $r_o = 1/g_o \approx 50$ kΩ.

These three values of g_i, g_m and g_o are entered into the equivalent circuit of Figure 4(e).

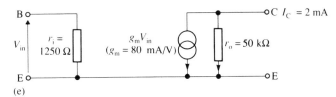

(e)

FIGURE 4 (e) a typical low-frequency small-signal equivalent circuit of a bipolar transistor at an operating point of $I_C = 2$ mA and $V_{CE} = 4$ V

☐   From an inspection of the graphs in Figure 4, how would you expect the input conductance, the transconductance and the output conductance of the transistor to change with increasing operating current?

■   Since the slopes of all the graphs in Figure 4 increase with increasing I_C all the conductances in the equivalent circuit must increase with increasing collector current.

Evidently graphical methods are neither a convenient nor a very accurate way of determining the equivalent circuit parameters. They do however give a useful visual picture of what the parameters mean and can give approximate values.

2.1.2 EQUIVALENT-CIRCUIT PARAMETERS DERIVED FROM NUMERICAL DATA

Manufacturers' data sheets usually give numerical values of certain small-signal parameters, though not always the equivalent-circuit parameter values. In addition, their data always refer to specific operating points at which the parameters were measured. This section explains how you can arrive at the small-signal equivalent-circuit parameter values at your chosen operating point.

(i) *Transconductance.* With bipolar transistors you do not need data sheets to tell you the value of g_m since it is always the same for well-made transistors. You can calculate its value from the equation for the collector current of a transistor given in Part 1 of this Block (i.e. the equation of the graph of Figure 4(b)). Since $I_C \approx I_E$

$$I_C \approx I_{SE} e^{KV_{BE}} \tag{1}$$

where K is about 40 V^{-1} at room temperature and I_{SE} is the emitter junction saturation current.

As before, the transconductance g_m is equal to dI_C/dV_{BE}, so

$$g_m = K I_{SE} e^{KV_{BE}}$$

or $$g_m \approx K I_C. \tag{2}$$

For example, if $I_C = 2$ mA then $g_m = 40\ V^{-1} \times 2$ mA $= 80$ mA/V, as found from the graph of Figure 4(b) in the last section.

Evidently this parameter does not depend on any quoted data: *the transconductance of all well-made bipolar transistors is the same. It is 40 V^{-1} times the d.c. operating collector current.*

☐ What is g_m when (a) $I_C = 10\ \mu A$, (b) $I_C = 0.25$ mA?

■ (a) When $I_C = 10\ \mu A$, $g_m = 40\ V^{-1} \times 10\ \mu A = 0.4$ mA/V.
(b) When $I_C = 0.25$ mA, $g_m = 40\ V^{-1} \times 0.25$ mA = 10 mA/V.

(Note that it is sometimes useful to refer to the emitter resistance r_e, which is the small signal resistance of the emitter-base p–n junction. $1/r_e = dI_E/dV_{BE}$ which differs only by α from dI_C/dV_{BE}, since α is about 0.99 it follows that $r_e \approx 1/g_m$.)

(ii) *Input conductance.* The input conductance is dI_B/dV_{BE} at the particular operating point. But since $I_B \approx I_C/\beta$ it follows that

$$g_i \approx g_m/\beta. \tag{3}$$

So if $\beta = 100$ and $I_C = 2$ mA

$$g_i = \frac{40\ V^{-1} \times 2\ mA}{100} = 800\ \mu S$$

or $\qquad r_i = 1250\ \Omega.$

(iii) *Output conductance.* The output conductance is determined by the Early effect, so it is necessary to know the Early voltage VA in order to calculate g_o.

As explained in Part 1, for a given V_{BE},

$$I_C \approx \text{constant} \times \left(1 + \frac{V_{CE}}{VA}\right): \tag{4}$$

So $\qquad g_o = \dfrac{dI_C}{dV_{CE}} \approx \dfrac{\text{constant}}{VA}.$

From equation (4) the 'constant' can be written as $I_C/(1 + V_{CE}/VA)$, so

$$g_o \approx \frac{I_C}{VA(1 + V_{CE}/VA)}$$

or $\qquad g_o \approx \dfrac{I_C}{VA + V_{CE}}. \tag{5}$

For example, if $VA = 100$ V, then, when $I_C = 2$ mA and $V_{CE} = 4$ V

$$g_o = 2\ mA/104\ V = 19.2\ \mu S$$

or $\qquad r_o = 1/g_o = 52\ k\Omega.$

This compares (rather better than might be expected) with the value obtained from the graph.

Note that g_i, g_o and g_m are all proportional to the operating collector current I_C and that g_o also depends on V_{CE}. So if you know their values at one operating point you can calculate their values at any other operating point.

SAQ 2
Use the above equations to derive the equivalent-circuit parameters of the transistor of Figure 4 at the operating point of $I_C = 1$ mA and $V_{CE} = 3$ V, given that β of the transistor is 100 and that the Early voltage is 150 V.

SAQ 3
What are the equivalent-circuit values for a transistor whose $\beta = 250$ and whose Early voltage = 100 V, when it is operating at $I_C = 0.1$ mA and $V_{CE} = 5$ V?

If transistor data sheets do not specify values of Early voltage, they usually give the output resistance r_o under specified operating conditions, from which the Early voltage can be calculated (using equation (5)).

SAQ 4	A transistor has the following small-signal equivalent-circuit parameters at an operating point of $I_C = 0.5$ mA and $V_{CE} = 5$ V: $r_i = 8000\ \Omega$, $r_o = 300$ kΩ and $g_m = 20$ mA/V. What values have the parameters β and VA, and what are the equivalent-circuit parameter values at an operating point of $I_C = 0.2$ mA and $V_{CE} = 10$ V?

Table 1 gives some parameter values which might appear in published data sheets for silicon n–p–n transistors. Note the very wide range of values for β.

Table 1 Some bipolar transistor parameters

Parameter	Test condition	Type A1	Type A2	Type A3
β_{min}	$V_{CE} = 5$ V, $I_C = 2$ mA	110	200	420
β_{max}		220	450	800
V_{BE}	$V_{CE} = 5$ V, $I_C = 2$ mA	0.55 V (min) to 0.7 V (max)		
f_T	$V_{CE} = 5$ V, $I_C = 10$ mA	150 MHz min		
h_{oe} (max)	$V_{CE} = 5$ V, $I_C = 2$ mA	30 μS	60 μS	100 μS
(typical)		15 μS	25 μS	45 μS

The parameter h_{oe} is one of four 'hybrid' parameters that are often used as an alternative way of specifying transistor small-signal performance. h_{fe}, the small-signal current gain already referred to, is another. The other two hybrid parameters are h_{ie} which is the input resistance, and h_{re} which is the small effect of changes in input voltage as a result of changes in output voltage. They are not discussed in this part of the course because each set of values applies to only one d.c. operating point, and it is not possible to calculate their values at other operating points except with the help of the kind of equivalent circuit discussed in this text.

The high-frequency parameter f_T referred to in Table 1 is approximately the frequency at which h_{fe} has fallen to 1. It indicates the highest frequency at which the transistor is likely to be useful as an amplifier (see Block 8).

The Early voltage is not given in the table, but the output conductance h_{oe} (which is virtually the same as g_o) at a given operating point, is given, from which you can calculate VA.

☐   What are the typical Early voltages of the three transistor types in Table 1?

■ $g_o = I_C/(VA + V_{CE})$; so for transistor Type A1

$$15\ \mu S = \frac{2\ \text{mA}}{VA + 5\ \text{V}}, \text{ so } VA = 128\ \text{V}.$$

Similarly for Type A2, $VA = 75$ V, and for Type A3, $VA \approx 40$ V. Notice that the larger the current gain the smaller is the Early voltage.

Summarizing for bipolar transistors:

The three small-signal equivalent-circuit parameters can be derived as follows:

(i) Transconductance g_m is always approximately KI_C. $K = 40$ V^{-1} at room temperature. g_m is the slope of the I_C/V_{BE} characteristic.

(ii) Input conductance $g_i = g_m/\beta$, and is the slope of the input characteristic.

(iii) Output conductance $g_o = I_C/(VA + V_{CE})$. It is the slope of the output characteristic when V_{BE} is held constant.

2.2   MOSFETS

2.2.1 EQUIVALENT-CIRCUIT PARAMETERS DERIVED FROM GRAPHS

Again the equivalent-circuit parameters can be derived from the graphs of the particular MOSFET's characteristics or from manufacturers' data. With MOSFETs, however, since the input conductance is zero (because there is no gate current) it is only necessary to pay attention to the output and transfer characteristics shown in Figures 5(a) and 5(b). The dashed line in Figure 5(a) marks the boundary between the *linear region of operation* and the *saturation region of operation.**

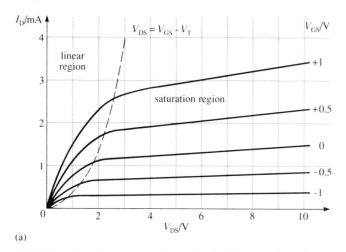

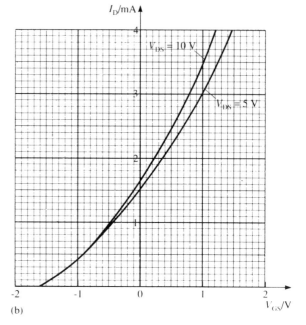

FIGURE 5 The d.c. characteristics of a depletion-mode n-channel MOSFET: (a) the output characteristics; (b) the forward transfer characteristics

SAQ 5 Estimate from the curves of Figure 5, (a) the output resistance, (b) the transconductance of the MOSFET at the operating point: $V_{DS} = 5$ V and $V_{GS} = +0.5$ V. Also note the threshold voltage.

2.2.2 EQUIVALENT-CIRCUIT PARAMETERS DERIVED FROM NUMERICAL DATA

Here again, measurements on graphs are not the most convenient or accurate way of obtaining equivalent-circuit parameters. But as with bipolars, their theoretical values can be calculated from the equations for the MOSFET output characteristics given in Part 1. These equations serve to identify which data are needed to determine the equivalent circuit parameters. The equations given in Part 1 are:

(a) *In the linear region where* $V_{DS} \leqslant V_{GS} - V_T$

$$I_D = \beta\{(V_{GS} - V_T) - V_{DS}/2\}V_{DS} \tag{6}$$

in which:

(i) β is the *gain factor* of the particular device. (It has no connection with the β of bipolar transistors!) Its value depends on the structure of the device as explained more fully in Part 3.

(ii) V_T is the *threshold voltage* (i.e. the gate voltage at which significant current starts to flow through the device. V_T is not easy to control precisely during manufacture so a spread of, say, 0.5 V to 4 V for an enhancement-mode device is not unusual.

(b) *In the saturation region, where* $V_{DS} \geqslant V_{GS} - V_T$

$$I_D = (\beta/2)(V_{GS} - V_T)^2(1 + \lambda V_{DS}) \tag{7}$$

---

* *The term 'saturation' here refers to the region of operation where the drain current tends to level off as* V_{DS} *increases. This is not the same use of the word as applied to diodes (i.e. 'saturation current'* I_S*), or as applied to bipolar transistors (i.e. 'saturation region of operation' where* V_{CE} *is less than* V_{BE} *as is fully explained in Part 3). The 'linear' region is so-called because at low values of* V_{DS} *the d.c. resistance between source and drain is linear (i.e. does not depend on* V_{DS}*) and is a function of* V_{GS}.

where λ is the channel length modulation factor. $1/\lambda$ is a voltage which is analogous to the Early voltage VA of a bipolar transistor.

(Note: You do not need to remember equations (6) and (7).)

Now, by differentiating equations (6) and (7), first with respect to V_{GS} and second with respect to V_{DS}, it is a simple matter to derive equations for the transconductance and the output conductance of a MOSFET in both the saturation region and the linear region of operation.

For example, in the *saturation region*, differentiating equation (7) with respect to V_{GS}, when V_T, λ and V_{DS} are constants, gives g_m. Thus

$$g_m = dI_D/dV_{GS}$$

so $\qquad g_m = \beta(V_{GS} - V_T)(1 + \lambda V_{DS}).$ $\hfill(8)$

But from equation (7) you can see that $(V_{GS} - V_T)^2 = 2I_D/[\beta(1 + \lambda V_{DS})]$. Substituting this value for $V_{GS} - V_T$ into equation (8) gives, after some manipulation

$$g_m = \sqrt{2\beta I_D(1 + \lambda V_{DS})} \hfill(9)$$
$$\approx \sqrt{2\beta I_D} \text{ when } \lambda V_{DS} \ll 1.$$

Note that the transconductance (in the saturation region) is proportional to the *square root* of I_D.

Similarly, differentiating equation (7) with respect to V_{DS}, when V_{GS}, V_T and λ are constants, gives the output conductance:

$$g_o = dI_D/dV_{DS} = \lambda(\beta/2)(V_{GS} - V_T)^2. \hfill(10)$$

But from equation (7) you can see that

$$(\beta/2)(V_{GS} - V_T)^2 = I_D/(1 + \lambda V_{DS}).$$

Substituting this value in equation (10) gives

$$g_o = \frac{\lambda I_D}{1 + \lambda V_{DS}} \hfill(11)$$
$$\approx \lambda I_D \text{ if } \lambda V_{DS} \ll 1.$$

It turns out, as explained later, that the maximum low-frequency voltage gain that a transistor can give can be expressed as $-g_m/g_o$. From equations (9) and (11) it is evident that, for a given MOSFET at a given drain voltage,

$$g_m/g_o \text{ is proportional to } 1/\sqrt{I_D} \text{ if } \lambda V_{DS} \ll 1.$$

Hence, in general, *the voltage gain that a particular MOSFET can give increases as the operating drain current is reduced.*

☐   The magnitude of the maximum voltage gain that a *bipolar* transistor can give can also be expressed as g_m/g_o. Does g_m/g_o in a bipolar depend on the d.c. operating collector current?

■   In a bipolar transistor $g_m = KI_C$ and $g_o \approx I_C/(V_{CE} + V_A)$ so g_m/g_o is independent of I_C.

Equations (6) to (11) can be used to calculate either the equivalent circuit parameters, g_m and g_o, or the performance parameters V_T, β and λ as follows.

Example: Using the characteristic curves of Figure 5 to obtain particular numerical data (instead of the slopes of graphs):

         (a) derive the performance parameters V_T, β and λ;

         (b) calculate the small signal equivalent circuit parameters g_m and g_o when $V_{DS} = 5$ V and $V_{GS} = 0.5$ V.

(a) *Calculating the performance parameters*

(i) From Figure 5(b) you can see that $V_T \approx -1.6$ V. Significant drain current only starts flowing once V_{GS} exceeds this voltage.

(ii) From a convenient point in the linear region in Figure 5(a), such as $I_D = 2.3$ mA and $V_{DS} = 2$ V when $V_{GS} = +1$ V, β can be calculated by substituting in equation (6), thus

since
$$I_D = \beta\{(V_{GS} - V_T) - V_{DS}/2\}V_{DS}$$

$$2.3 \text{ mA} = \beta(1 \text{ V} + 1.6 \text{ V} - 1 \text{ V})2 \text{ V}$$

$$= \beta \times 1.6 \text{ V} \times 2 \text{ V} = 3.2\beta \text{ V}^2.$$

Hence
$$\beta = \frac{2.3 \text{ mA}}{3.2 \text{ V}^2} = 0.72 \text{ mA V}^{-2}.$$

(iii) The d.c. characteristics, when extrapolated 'backwards', meet the voltage axis at a voltage of $-1/\lambda$. In this case $1/\lambda \approx 30$ V so $\lambda \approx 0.033$ V^{-1}.

(b) *Calculating the equivalent-circuit parameters*

First find the d.c. operating current.

From Figure 5(b), when $V_{DS} = 5$ V and $V_{GS} = 0.5$ V you can see that $I_D \approx 2.1$ mA.

(iv) Substituting this value in equation (11) gives g_o. Therefore

$$g_o = \frac{0.033 \text{ V}^{-1} \times 2.1 \text{ mA}}{1 + 0.033 \text{ V}^{-1} \times 5 \text{ V}} = 59 \text{ μS}$$

or $r_o \approx 17$ kΩ.

(v) At the same operating point equation (8) gives

$$g_m = \beta(V_{GS} - V_T)(1 + \lambda V_{DS})$$

$$= 0.72 \times 2.1 \times 1.165 \text{ mA/V} \approx 1.8 \text{ mA/V}.$$

| SAQ 6 |

(a) Using equation (6) derive equations for the transconductance and the output conductance of a MOSFET in the linear region of operation.

(b) From the performance parameters derived from Figure 5, and the equations you have just derived, calculate values for g_m and g_o for the transistor at the operating point of $V_{DS} = 1$ V and $I_D = 1.5$mA.

Summarizing for MOSFETs

(i) Transconductance: (a) in the saturation region $g_m \approx \sqrt{2\beta I_D}$; (b) in the linear region $g_m = \beta V_{DS}$.

(ii) Output conductance: (a) in the saturation region $g_o \approx \lambda I_D$; (b) in the linear region $g_o = \beta(V_{GS} - V_T - V_{DS})$.

So the performance parameters needed in order to calculate the equivalent circuit parameters at any operating point are β, V_T and λ.

Table 2 shows a manufacturer's data for a type of discrete MOSFET (i.e. one that is not part of an integrated circuit). From these data, because the operating points are given, it is possible to calculate the three key performance parameters.

Table 2 Some data for an n-channel depletion-mode MOSFET

Parameter	Test condition	Type number BFR 29
g_m	$V_{DS} = 15$ V, $I_D = 2$ mA	6 mA/V (min)
I_{DSS}	$V_{DS} = 15$ V, $V_{GS} = 0$	10 to 40 mA
$V_T(I_D \approx 0)$	$V_{DS} = 15$ V	-4 V (max); range: -0.5 V to -4 V
g_o	$V_{DS} = 15$ V, $I_D = 2$ mA	400 μS (max)

Remember that I_{DSS} is the current which flows through the MOSFET when $V_{GS} = 0$ (see Section 5.2 of Part 1).

The magnitude of the maximum voltage gain obtainable, namely g_m/g_o, with this transistor, at the operating point of $V_{DS} = 15$ V and $I_D = 2$ mA is 6 mA/V ÷ 400 μS = 15. (This is the lower limit; it will be greater for a typical transistor and at a lower operating current.) For a typical bipolar transistor at the same operating point the gain available is at least 80 mA/V ÷ 30 μS $\approx$ 2600. Hence the preference, on the whole, for bipolar transistors in op amps!

3 BASIC TRANSISTOR CIRCUITS

In the following subsections four basic circuits are described and explained. There are several possible variations of each circuit which have different properties, as I shall explain. With each circuit I shall first show how to set the operating point of the circuit; I shall then show how to draw the small-signal equivalent circuit of the circuit, from which it is possible to calculate the circuit's voltage gain, input resistance and output resistance; and I shall finally show how, by understanding how the circuit works, you can quite simply arrive at quite good estimates of these performance data.

3.1 THE COMMON-EMITTER AMPLIFIER

3.1.1 THE BASIC AMPLIFIER CIRCUIT

The d.c. components of the circuit of a simple common-emitter amplifier are shown in Figure 6(a). To amplify, this circuit is connected to a signal

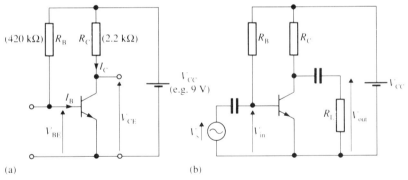

(a) (b)

FIGURE 6 The simple common-emitter amplifier circuit with components appropriate for a transistor whose β is 100, operating at a collector current of 2 mA: (a) the d.c. circuit arrangement; (b) the d.c. circuit coupled via capacitors to the source and load (the capacitors are to ensure that the d.c. levels are not disturbed by connecting the external voltage source and load).

voltage source and to a load R_L by capacitors as shown in Figure 6(b). The first task is to choose the resistors in the d.c. circuit so that a suitable operating point is set up; that is, to establish appropriate and well-defined values of V_{CE}, I_C, I_B and V_{BE}. (Remember that these capital symbols with *capital* subscripts refer to *d.c. values*. Capital symbols with *lower case* subscripts usually refer to *amplitudes* of sinusoids.)

The d.c. circuit design: Figure 6(a)

(i) *Collector-emitter voltage V_{CE}.* To allow the collector voltage to vary about its operating point in response to the input signal, an operating value of V_{CE} of about half V_{CC} is usually chosen: about 4.5 V in this example. This allows the largest possible output signal amplitude on either side of the operating point without 'clipping' occurring. If the d.c. operating point is set too high the 'top' of the waveform can be cut off if it exceeds V_{CC}. Similarly the 'bottom' can be cut off if the d.c. operating point is too low. Obviously both 'ends' are clipped if the signal amplitude is too large.

(ii) *Collector current I_C.* The operating collector *current* to choose usually depends on the output required of the circuit. In this example there are no output requirements specified so the choice is an arbitrary one. A d.c. collector current of 2 mA has been chosen (though any current from 10 μA to 10 mA might be appropriate for a small transistor.) *This fixes the value of the collector resistance R_C at 2.2 kΩ* (i.e. the 4.5 V voltage drop across it divided by 2 mA).

(iii) *Base current I_B.* To obtain the specified collector current, an appropriate input to the base must be provided. In Figure 6(a) a well-defined base current is provided through R_B, which results in the usual V_{BE} value of about 0.65 V appearing across the emitter-base p–n junction as explained in Part 1. *The base current needed depends on the current gain β of the transistor since $I_B = I_C/β$.* So if $β = 100$ then in this case $I_B = 2\text{ mA}/100 = 20$ μA.

Evidently, in Figure 6(a)

$$I_B = \frac{V_{CC} - V_{BE}}{R_B}.$$

Since V_{BE} varies by only a few tens of millivolts from one transistor to another, it is usually accurate enough to assume a value of, say, 0.65 V for V_{BE} in this equation. So in this case

$$20\ \mu A \approx \frac{9\text{ V} - 0.65\text{ V}}{R_B}$$

giving $\qquad\qquad R_B \approx 420\text{ k}\Omega.$

Thus the d.c. operating point of the transistor in this circuit is established as $I_C = 2$ mA, $V_{CE} \approx 4.5$ V, $I_B = 20$ μA and $V_{BE} \approx 0.65$ V.

Note that the range of possible values of β—called the **spread** of β—is very wide, even for a given type of transistor since it is difficult to control this value in manufacture. Typically β may be anything from 100 to 400 or more, so designing for a particular value of β is only satisfactory if the transistors to be used are very carefully selected for their β values. One of the main problems with bipolar transistor circuit design is making allowance for the wide spread of β values. A similar problem arises with MOSFETs, where the spread is in values of threshold voltage and I_{DSS}.

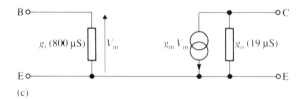

(c)

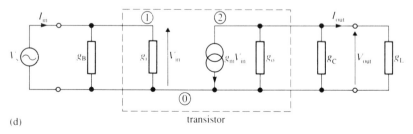

(d)

FIGURE 6 (c) the corresponding transistor small-signal equivalent circuit; (d) the equivalent circuit of the complete amplifier, including the transistor

SAQ 7 Figure 6(c) shows the small-signal equivalent-circuit of the transistor in Figure 6(a). Given that the Early voltage of the transistor is 100 V, confirm that the equivalent-circuit parameter values given in Figure 6(c) are correct for the operating point of the circuit of Figure 6(a).

The small-signal equivalent circuit of the common-emitter amplifier

Figure 6(b) shows the transistor coupled to a source and load by capacitors so that the d.c. levels will not be disturbed. To calculate the amplifier's small-signal performance using nodal analysis the equivalent circuit of the *transistor* is inserted into the circuit of the amplifier in place of the transistor symbol. The battery is also replaced in the circuit diagram by its small-signal equivalent circuit, namely a short circuit, thus forming Figure 6(d). (Note that in this illustration the capacitors have been assumed to have negligible reactance at the frequency of operation and have therefore been ignored.) For nodal analysis all resistances are best shown as conductances, as shown in the figure.

☐   Why is the small-signal equivalent-circuit of a battery usually taken to be a short circuit?

■   Because ideally there is zero change of voltage across the battery when the current through it changes. That is, $dV/dI = 0$, so the equivalent small-signal resistance is zero. In practice the 'battery' is usually a mains-derived d.c. supply, and there is usually a large capacitor across the output of it so that it presents a very low a.c. impedance to the circuit it is supplying.

The effect of regarding the supply as a small-signal short circuit is that $g_B(= 1/R_B)$ appears in parallel with g_i of the transistor equivalent circuit, and that $g_C(= 1/R_C)$ appears in parallel with g_o as shown in Figure 6(d).

When you use a computer to calculate the circuit's performance you can enter all the component values in Figure 6(d) into the computer net list and let the computer solve the equations; but most computer simulation programs allow you to skip the stage of drawing the small-signal equivalent circuit. You simply enter the resistor and generator values in Figure 6(b) into the net list, and call up the transistor type you are using from the 'library' in the computer. The computer then calculates the d.c. operating point, the transistor equivalent circuit values and the voltage gain, input resistance etc. (see Computing Exercise 1).

With this circuit however the nodal analysis turns out to be very straightforward, so I will carry out the performance calculations without the aid of the computer.

Estimating the small-signal performance of the amplifier

Assuming that the transistor is as specified in Figure 6(c), namely $g_i = 800\ \mu S$, $g_o \approx 19\ \mu S$ and $g_m = 80\ mA/V$, the circuit performance can be estimated as follows:

(i) *open-circuit voltage gain*. The term 'open circuit' refers to the 'no load' condition, implying that $I_{out} = 0$, (i.e. $R_L = \infty$ or $g_L = 0$) so for this calculation the load conductance is omitted from the diagram. The current equation at node 2 in Figure 6(d) becomes

$$g_m V_{in} + V_{out} g_o + V_{out} g_C = 0$$

so $$A_v = \frac{V_{out}}{V_{in}} = -\frac{g_m}{g_o + g_C}$$

$$= -\frac{80\ mA/V}{19\ \mu S + 455\ \mu S}$$

$$= -169.$$

The minus sign indicates that the circuit inverts the signal or produces a 180° phase change in a sinusoidal waveform. Notice that the conductance g_C of R_C is much greater than the output conductance of the transistor, and therefore dominates the denominator of the above equation. Higher gains should therefore be possible if g_C could be made much smaller.

With the external load R_L connected, a conductance g_L appears in the equivalent circuit in parallel with g_o and g_C, making a net load conductance of $g_o + g_C + g_L$, and giving

$$A_v = -\frac{g_m}{g_o + g_C + g_L}. \qquad (12)$$

This illustrates a general point that it is helpful to remember: *the voltage gain A_v of such a circuit is $-g_m \times R_{L\,(total)}$, where $R_{L\,(total)}$ is the resistance of all the resistances connected in parallel between the output terminal and ground.*

Note that if g_L and g_C can be made small enough to be neglected, the voltage gain is the maximum that a particular device can give. Thus

$$A_{v\,(max)} = -g_m/g_o.$$

This is the value given earlier for the maximum voltage gain that a transistor can give.

(ii) *Output resistance with short-circuit input.* The output resistance of the amplifier is the resistance that the load 'sees' looking back into the circuit when the current generator in the equivalent circuit is inactive. At node 2 again

$$V_{out}g_o + V_{out}g_C = -I_{out}$$

so output resistance R_{out} is given by

$$R_{out} = \frac{V_{out}}{-I_{out}} = \frac{1}{g_o + g_C}. \qquad (13)$$

In this example,

$$R_{out} = \frac{1}{19\ \mu S + 455\ \mu S} = 2.11\ k\Omega.$$

But how is the current generator $g_m V_{in}$ made inactive? It is only inactive if there is no signal voltage at the input, which means that the d.c. input voltage must have no signal superimposed on it. This is referred to as **short-circuit input** (as far as signals are concerned). If any signal is present at the input then clearly the output current will not be due solely to g_C and g_o. This is why it is necessary to specify 'short-circuit input' if equation (13) is to apply.

Note that once again the conductance of R_C dominates the expression.

(iii) *Input resistance.* At node 1, equating currents gives

$$I_{in} = V_{in}(g_B + g_i)$$

so $$R_{in} = \frac{V_{in}}{I_{in}} = \frac{1}{g_B + g_i}. \qquad (14)$$

With $R_B = 420\ k\Omega$, g_B can almost be ignored, so $R_{in} \approx 1/g_i = 1250\ \Omega$.

Note that the *open-circuit* voltage gain V_{out}/V_{in} is not affected by the capacitors shown in Figure 6(b) even if their reactances are not negligible, because V_{out} and V_{in} are measured at the base and collector terminals. If the capacitor reactances are not negligible $V_{in} < V_s$, and with the load connected V_{out} is reduced. The capacitors produce a 'single lead' low frequency response of the circuit as a whole. (The frequency dependent parameters of the *transistor*, which have not yet been discussed, produce a 'single lag' response at high frequencies as explained in Block 8.)

> **SAQ 8**
>
> The load resistance R_L in Figure 6(b) is 8 kΩ. The input is driven by a source whose open-circuit output voltage amplitude is 5 mV and whose internal resistance is 1 kΩ. The coupling capacitors have

negligible reactances. Draw the small-signal equivalent circuit of this arrangement and calculate the amplitude of the output signal voltage across the load. (First calculate V_{in} by regarding the input circuit as a potential divider driven by V_s. Then calculate the voltage gain V_{out}/V_{in}.)

A quicker but more approximate estimate of circuit performance runs as follows. (i) Since R_B can be neglected when in parallel with g_i, the input resistance of the amplifier is approximately $1/g_i$ (1250 Ω). (ii) Since g_o can be neglected in comparison with $1/R_C$, $R_{out} \approx R_C$. The open circuit voltage gain is therefore $-g_m R_C$. Thus in this case

$$A_v \approx - 80 \text{ mA/V} \times 2.2 \text{ k}\Omega = -176.$$

Exercise 1 in the Home Computing Booklet for Block 4 enables you to simulate the performance of this amplifier using NODALOU.

3.1.2 PRACTICAL COMMON-EMITTER AMPLIFIER DESIGNS

In this section I want to consider the performance of circuits which have better d.c. designs than the circuit of Figure 6(a).

SAQ 9 Explain why the circuit of Figure 6(a) is in general unsatisfactory, by considering the operating point of the circuit when $\beta = 200$ or more, instead of 100.

Figures 7(a) and (b) show two ways of reducing the variation of operating point as a result of differences in the values of β.

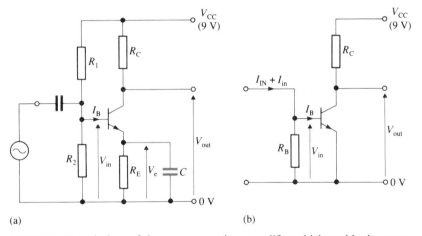

(a)   (b)

FIGURE 7 Two designs of the common-emitter amplifier which enable the operating point of the transistor to be better defined despite variations of β: (a) with a feedback resistor R_E. The capacitor shown in grey restores the small-signal gain without upsetting the d.c. operating point; (b) a d.c. coupled circuit of the kind used in op amps. The control of the operating point would be achieved mainly by overall d.c. feedback around the op amp of which this circuit forms a part

In Figure 6(a) the resistor R_B established the base current of the transistor. In Figure 7(a) R_B is replaced by resistors R_1 and R_2 which form a potential divider and set the *voltage* level of the base terminal. The emitter current is then determined mainly by the emitter resistor R_E.

For example, if you want to design a circuit in which the operating emitter current is 2 mA you can choose R_E to be 500 Ω to produce a voltage drop across it of 1 V. Since $V_{BE} \approx 0.65$ V it follows that the base terminal should be held at a voltage of about 1.65 V. With $V_{CC} = 9$ V, and if I_B is small compared with the current flowing in R_1 and R_2 for all likely values of β, this can be achieved with the potential divider by making the ratio $R_2/R_1 = 1.65/7.35$ (e.g. $R_2 = 1.65$ kΩ and $R_1 = 7.35$ kΩ or $R_2 = 16.5$ kΩ and $R_1 = 73.5$ kΩ). Then I_E and I_C will not deviate much from 2 mA even for extreme β values of 100 or 400.

☐   Which pair of values of R_1 and R_2 suggested in the above paragraph will give a better control of I_C as β varies from one transistor to another?

■   The spread of β is reflected in the spread of I_B. For these differences in I_B to have minimal effect on the base voltage, R_1 and R_2 should be as small as possible. So the lower of the two pairs of resistances will give the better control of I_E. However, the smaller these resistances are, the more d.c. current will flow straight through them to the zero voltage line and be wasted in heat generation, and the smaller will be the circuit's input resistance, so there is a practical limit to how small these resistors should be.

Estimating the gain of the circuit of Figure 7(a)

The voltage gain of the circuit can be estimated approximately as follows (assuming that g_o of the transistor is negligibly small compared with g_C).

The signal currents through R_C and R_E are approximately the same ($I_c \approx \alpha I_e$), so the voltage drops, V_{out} and V_e, across them are in proportion to their resistances. That is $V_{out}/R_C = -V_e/R_E$ (the minus sign indicating $180°$ phase difference). And since there is relatively little *signal* voltage drop across the emitter p–n junction, $V_{in} \approx V_e$. Therefore

$$\frac{V_{out}}{R_C} \approx -\frac{V_{in}}{R_E} \text{ or } A_v = \frac{V_{out}}{V_{in}} \approx -\frac{R_C}{R_E}.$$

☐   If $I_E = 2$ mA and if R_C and R_E are, respectively, 2.2 kΩ and 500 Ω, what is the voltage gain of the circuit?

■   From the above equation $A_v \approx -2.2 \text{ kΩ}/500 \text{ Ω} = -4.4$.

Evidently the inclusion of R_E in the circuit reduces the voltage gain considerably.

| SAQ 10 | That the gain of the circuit of Figure 7(a) is the ratio of two resistances, and that the variations of β have little effect on the gain suggest that the circuit is a negative feedback one. Explain the operation of this circuit from a feedback point of view. Is it voltage or current derived, and is the feedback shunt or series connected? How does this influence the input resistance of the circuit? |

A rather more accurate estimate of voltage gain takes into account the fact that the current through the transistor also flows through the emitter resistance r_e, causing V_b to be somewhat bigger than V_e. In other words, r_e must be added to R_E to give a better estimate of V_{in}. But, as pointed out in Section 2.1.2, $r_e \approx 1/g_m$, so

$$A_v \approx -\frac{R_C}{R_E + 1/g_m}. \tag{15}$$

If again $I_E = 2$ mA, $r_e = 12.5$ Ω. So, by equation (15), when $R_C = 2.2$ kΩ and $R_E = 500$ Ω

$$A_v = -\frac{2.2 \text{ kΩ}}{500 \text{ Ω} + 12.5 \text{ Ω}} = -4.3.$$

The gain can, however, be restored by the inclusion of capacitor C, which appears grey in the figure (see Exercise 2 of the Computing Booklet for this block). This does not disturb the d.c. operating point, but if the capacitor has a very small reactance at the frequency of operation, the signal bypasses R_E so that R_E is virtually shorted out as far as the signal is concerned. The gain of the circuit therefore becomes about the same as that of Figure 6(a).

| SAQ 11 | How would you modify the a.c. circuit of Figure 7(a), including the capacitor, again without altering the d.c. operating point, in such a way as to achieve a voltage gain of between $-R_C/r_e$ and $-R_C/(R_E + r_e)$; a voltage gain of -22, for example? |

The very large capacitor needed in the circuit of Figure 7(a) to obtain a large gain is very inconvenient for some transistor circuits and is quite impossible in integrated circuits in which capacitors of only a few picofarads are possible. The circuit of Figure 7(b) is often used instead.

The circuit of Figure 7 (b) is called a **direct-coupled** circuit. The input signal current I_{in} is supplied along the same wire as the d.c. bias current I_{IN}. Both currents divide at the base terminal, some flowing through R_B and some flowing through the transistor. R_B is chosen so that, for a typical transistor, there is a voltage drop across it, due to $(I_{IN} - I_B)$, of about 0.65 V, as required to bias the transistor properly. It turns out that the variation in I_C as a result of the variation in β is reduced. The greater you make I_{IN}, and the smaller you make R_B, the smaller the variations in I_C. The main advantage of the circuit of Figure 7(b), however, arises from the possibility of d.c. coupling it to the circuits preceding it and following it. When the circuit is included in a high-gain amplifier (such as an op amp) to which overall negative feedback is being applied, both the small-signal performance *and* the d.c. behaviour of the circuit are controlled mainly by the feedback. The d.c. operating points acquire approximately the intended d.c. levels despite the effects of variations in the β values of the transistors.

The small-signal equivalent circuit of Figure 7(b) is the same as that of Figure 6(a), so the open circuit voltage gain is also the same: about -169.

☐   *Revision*. What other normally desirable effects does negative feedback applied to a high-gain amplifier provide?

■   Negative feedback arrangements also diminish the effects of temperature changes, diminish the effect of differences in other parameters such as r_o, reduce the interference due to noise generated within the transistor and reduce the distortion produced by the transistor's nonlinear characteristics.

3.2   THE USE OF A DYNAMIC LOAD

As already noted, the voltage gain of a common-emitter amplifier is given by $g_m \times R_{L(total)}$, where $R_{L(total)}$ consists of the resistance of R_C in parallel with both the external load resistance and the output resistance of the transistor. In the circuits of Figure 6(a), 7(a) and (b), open-circuit output conditions were considered ($R_L = \infty$) and R_C was very much less than r_o so that the gain of the circuit was approximately $g_m R_C$ (i.e. r_o could be neglected). But evidently, if R_C were to be increased, higher values of open-circuit gain should be possible.

☐   Why is it not possible to increase the gain of Figure 6(a) significantly simply by using a higher resistance resistor at R_C?

■   Suppose you double the resistance of R_C. The operating current I_C of the transistor must be halved (or thereabouts) in order that the operating voltage will remain at about $V_{CC}/2$. But if I_C is halved then g_m is also halved; so the voltage gain ($g_m R_C$) remains about the same. Note however that the gain can be varied somewhat by changing the value of R_C if the consequent effect on the d.c. operating collector voltage is acceptable.

A circuit that can supply the d.c. current that the collector requires, at the same time as presenting a large small-signal resistance, is the *current source* described in Part 1. Figure 8(a) (overleaf) shows a current source in the form of a current mirror connected to the amplifying transistor in place of the resistor R_C. In this application the current mirror is referred to as a **dynamic load**.

3.2.1 THE D.C. DESIGN

You should recall from Part 1 that a current mirror works as follows. Referring to the circuit in Figure 8(a), in which the current mirror is made from p–n–p transistors, the voltage drop across T2 is the usual 0.65 V of a forward-biased p–n junction, so the current I_2 flowing through it and through R_1 is about (18 V–0.65 V)/R_1 assuming +9 V and –9 V supplies. The voltage drop across the base-emitter of T2 is also the base-emitter voltage of T3, so because the two

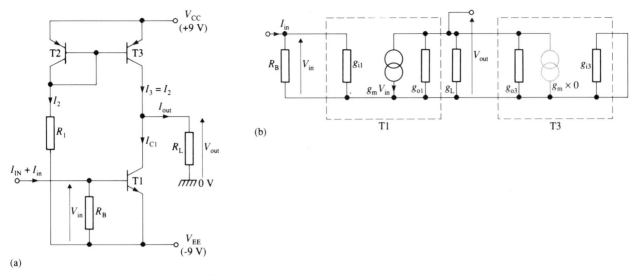

(a)

(b)

FIGURE 8 (a) A common-emitter amplifier with 'dynamic load' to increase its voltage gain; (b) the small-signal equivalent circuit of T1 and T3

transistors are virtually identical, the collector current I_3 of T3 must be equal to I_2.

If in addition I_{IN} and R_B are chosen so that the collector current I_{C1} is equal to I_3, then T1 and T3 can be thought of as two equal d.c. current sources facing each other, one receiving the current that the other supplies, so that there is no current I_{OUT} available to flow in the load and there is no voltage drop across R_L. The d.c. operating *voltage* of the collectors of T1 and T3 is fixed by the 'reference voltage' to which R_L is connected—which is 0 V in this case. Thus $V_{OUT} = 0$.

When however a signal voltage is applied to the input of T1 the collector current of T1 varies but that in T3 does not, so the amplified signal current is driven into the load causing a signal voltage to appear across it.

☐   What are the maximum and minimum possible values of the output voltage in Figure 8(a) before clipping starts?

■   Transistors T1 and T3 will go on behaving properly as amplifiers so long as the magnitude of V_{CE} of each transistor does not fall much below V_{BE}. Since V_{BE} of each transistor is about 0.65 V as usual, the output voltage can therefore reach a maximum of about $(9\,V - 0.65\,V) = 8.35\,V$ and a minimum of about $-8.35\,V$.

3.2.2 THE SMALL-SIGNAL EQUIVALENT CIRCUIT

The small-signal equivalent circuit of Figure 8(a) is shown in Figure 8(b). Only T1 and T3 are included in the equivalent circuit because T2 is simply being used as a constant low-resistance d.c. voltage source and there is no signal current flowing in it. The current generator in T3 is therefore inactive and appears grey to indicate that it can be ignored. In other words T3 is simply acting as a collector resistance for T1, whose resistance is the output resistance of T3. This circuit can be analysed to give its small-signal performance using your computer (see Exercise 2 in the Computing Booklet).

3.2.3 ESTIMATING THE VOLTAGE GAIN OF THE COMMON-EMITTER AMPLIFIER WITH DYNAMIC LOAD

Looking at the central part of figure 8(b) you can see that the output conductances g_{o1} and g_{o3} of T1 and T3 and the conductance g_L of the load resistance are all in parallel at the output node; so from equation (12),

$$\frac{V_{out}}{V_{in}} = -\frac{g_m}{g_{o1} + g_{o3} + g_L} \tag{16}$$

Under open-circuit output conditions in which $g_L = 0$, and assuming that $g_{o1} = g_{o3} = 19\ \mu S$ as before when $I_C = 2\ mA$,

$$A_v = \frac{-0.08\ \text{A/V}}{2 \times 19\ \mu S} = -2105.$$

The voltage gain has been increased by about an order of magnitude simply by increasing the net load resistance $R_{L\,(total)}$.

The output resistance is, however, now very high (about $1/2\,g_o$), so it is essential, if advantage is to be taken of this increased gain, to connect an external load whose resistance is at least of the same order of magnitude as the transistors' output resistances. (An emitter-follower (see Section 3.4) is often therefore used as the load for the circuit of Figure 8(a). Indeed it is possible to use an emitter-follower *in place of* the current mirror, as explained in Section 4.)

The input resistance of the amplifier is little affected by these alterations to the output circuit.

□   If the Early voltage of both T3 and T2 is 90 V and their collector voltages are 4 V, what are their output resistances, and what is the open-circuit voltage gain of the circuit?

■   The output resistance of a bipolar transistor is $(VA + V_{CE})/I_C$ (see equation (5)). So in this case the output resistances are each $94\ V/2\ mA = 47\ k\Omega$. The voltage gain of the circuit is therefore $-g_m R_{L\,(total)} = -80\ \text{mA/V} \times 47\ k\Omega/2 = -1880$.

> **SAQ 12**   In the circuit of Figure 8(a), the operating currents of all three transistors are set at 1 mA. The operating collector current of T1 is set by adjusting I_{IN}. The current through the current mirror is set by R_1. The Early voltage of all three transistors is 150 V. Calculate what the resistance of R_1 should be, and calculate the voltage gain of the circuit if (a) R_L has a resistance of 500 kΩ; (b) R_L has a resistance of 5 kΩ.

3.3   THE LONG-TAILED PAIR

The **long-tailed pair** provides one of the most effective ways of establishing the operating point of an amplifier. The circuit is shown in Figure 9(a). A current source supplies the d.c. current for *two* transistors whose emitters are connected together as shown.

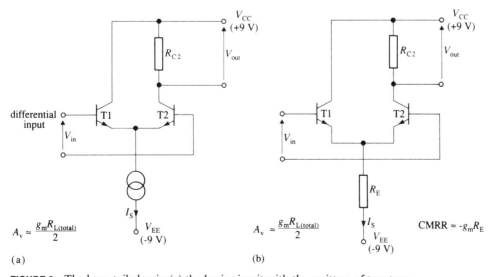

FIGURE 9   The long-tailed pair: (a) the basic circuit with the emitters of two transistors connected to a single current source; (b) the original form of the basic circuit with a resistor R_E in place of the current source

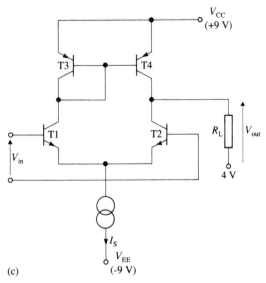

(c)

FIGURE 9 (c) the basic circuit with a p–n–p current mirror acting as a dynamic load

The circuit works as follows. The signal voltage is applied between the base terminals of the two transistors, and is therefore a *differential* input voltage, as in an op amp. When the differential input voltage is zero the two transistors each take half of the constant current I_S which is supplied by the current source. But when a small differential voltage v_{in} is applied to the input terminals, the current through one transistor increases and the current through the other decreases, leaving the total current unaltered. The change in current in T2 causes a change in the voltage drop across R_{C2} thus producing an output voltage v_{out}, and if v_{out} is greater than v_{in} the circuit is a voltage amplifier.

☐ If a resistor R_{C1} of the same resistance as that of R_{C2} were inserted into the collector lead of T1 a second output voltage could be obtained. How would this differ from the output from T2?

■ It would be the inverse of the output from T2; that is, it would be 180 degrees out of phase, but of the same amplitude. This is because the current through T1 increases by the same amount as the current through T2 decreases—and vice versa. As you might expect, the two outputs considered together are called a 'differential output'.

The circuit has a number of advantages over the common-emitter amplifier. Firstly, since it has a *differential input* like an op amp, it is not essential for one side of the signal source to be connected to the common, zero voltage line. Secondly the d.c. voltage levels of the two inputs can vary *together*, without affecting the output very much. That is, a coupling capacitor is not needed to isolate changing or different d.c. levels associated with the input signal, as in the common-emitter amplifier. Thirdly variations in temperature do not affect the circuit much because they affect both transistors equally.

3.3.1 THE D.C. DESIGN

The d.c. design is very straightforward. The operating current of each transistor is half the current supplied by the current source, namely $I_S/2$, and the collector voltage of T2 is clearly $V_{CC} - I_S R_{C2}/2$. That's almost all there is to it!

Because the d.c. voltage of the input terminals can vary a good deal without affecting the circuit performance very much, the operating collector voltages of the transistors can be expected to vary over a similar range.

☐   If the voltage drop across R_{C2} in Figure 9(a) is designed to be 2 V, and if the current source requires 1 V across it if it is to work properly, what is the possible range of d.c. input voltage?

■   Transistor T2 will still work satisfactorily with small signals even if $V_{CB} = 0$ V; so the input voltage can rise to $+7$ V (i.e. 9 V minus the voltage drop across R_{C2}). If there is to be 1 V across the current source, the d.c. input voltage must be about 0.65 V more positive than this in order to provide the base-emitter voltage for T1 and T2. So the lowest possible input voltage is about -7.35 V.

3.3.2 ESTIMATING THE PERFORMANCE OF THE LONG-TAILED PAIR

Since you can learn much more about the long-tailed pair by studying how it works than by nodal analysis of it, I have left consideration of the equivalent circuit till last.

You can think of the long-tailed pair as consisting of two common-emitter amplifiers which are sharing the one input voltage. Each transistor receives half the input voltage V_{in} because, for small signals, the voltage of the node joining the two emitters remains midway between the voltages of the two input terminals, so the net gain of T2 is half that of the normal common-emitter circuit. Hence the differential voltage gain becomes

$$A_{v\,(diff)} = \frac{V_{out}}{V_{in}} \approx \frac{g_m R_{L\,(total)}}{2} \qquad (17)$$

where $R_{L\,(total)}$ as before is the net load resistance of T2 and includes its own output resistance. Notice that the output of T2 is not inverted with respect to the input.

☐   If $I_S = 200$ µA and the d.c. voltage drop across R_{C2} is 2 V, what is the open-circuit differential voltage gain of the circuit, assuming that r_o of T2 can be neglected?

■   The collector current of T2 is 100 µA, so $g_m = 4$ mA/V and $R_{C2} = 20$ kΩ. Therefore

$$A_{v\,(diff)} = \frac{4\ \text{mA/V} \times 20\ \text{k}\Omega}{2} = 40.$$

The differential input resistance is evidently the input resistances of the two transistors in series; so $R_{in\,(diff)} = 2r_i$. Thus if β of each transistor is 200, and the transconductance is 4 mA/V (as in the in-text questions above), then

$$R_{in\,(diff)} = \frac{2\beta}{g_m} = \frac{400}{4\ \text{mA/V}} = 100\ \text{k}\Omega.$$

As with the common-emitter amplifier, the resistance of the collector resistor R_{C2} is normally much less than r_o of T2, so $R_{out} \approx R_{C2}$.

3.3.3 COMMON-MODE REJECTION

The original design of the long-tailed pair has a resistor R_E as the 'long tail', as shown in Figure 9(b) on p. 67, instead of the current source. This circuit suffers considerably from a low value of an important additional parameter known as the **common-mode rejection ratio** (CMRR). This is the ratio of the differential gain, just discussed, to the **common-mode gain**, which can be explained as follows.

If you connect the two inputs together, so that there cannot be a *differential* input voltage, and then apply a positive voltage ΔV_{IN} to *both* inputs, it is evident that the voltage across R_E and therefore the current through it will be increased, and that this will cause a change ΔV_{OUT} in the output voltage. The ratio of these two voltages, $\Delta V_{OUT}/\Delta V_{IN}$, is the common-mode gain. Its value can be estimated as follows.

If ΔI_E is the increase in current through R_E caused by an increase ΔV_{IN} in the common-mode input, then $\Delta I_E/2$ is the increase in current through

R_{C2}. The decrease ΔV_{OUT} in the output voltage is therefore $(\Delta I_E \times R_{C2})/2$ (assuming that r_o of T2 is much larger than R_{C2}).

But ΔI_E is evidently $\Delta V_{IN}/R_E$, so

$$\Delta V_{OUT} \approx -\frac{\Delta V_{IN} R_{C2}}{2R_E}$$

and the

$$\text{common-mode gain} = \frac{\Delta V_{OUT}}{\Delta V_{IN}} \approx -\frac{R_{C2}}{2R_E}. \tag{18}$$

Since the

$$\text{differential voltage gain} \approx \frac{g_m R_{C2}}{2}$$

it follows that the

$$\text{CMRR} = \frac{\text{differential gain}}{\text{common-mode gain}} \approx \frac{g_m R_{C2}}{2} \div -\frac{R_{C2}}{2R_E} = -g_m R_E. \tag{19}$$

The use of a current source in place of R_E, as in Figure 9(a), evidently improves the CMRR because the output resistance of the current source is much greater than the resistance of R_E.

That is $\text{CMRR} = -g_m \times$ (the output resistance of the current source).

A large CMRR is important because it diminishes the effect of unwanted common-mode inputs on the differential gain. One troublesome common-mode input is that due to temperature changes. The base-emitter voltage of a transistor decreases by about 2 mV per degree centigrade. Thus a 1°C temperature rise of transistors T1 and T2 is equivalent to a *common-mode* input voltage of 2 mV, and it is obviously desirable that this should not affect the normal output too much. The CMRR is a measure of how well the amplifier achieves this rejection of common-mode signals. A CMRR of better than 10^5 or 100 dB is a typical value for an op amp (see Exercise 3 in the Home Computing Booklet for Block 4).

SAQ 13

(a) In Figure 9(b) if $I_S = 200\ \mu A$, and the d.c. input voltage can be expected to be around zero volts, what will be the CMRR?

(b) Redraw the circuit diagram of Figure 9(a) to include the transistors in the current mirror which forms the current source. For $V_{CC} = -V_{EE} = 9$ V determine the appropriate resistor value to give $I_S = 200\ \mu A$. If the required CMRR is -3000, what value of Early voltage should the output transistor of the current source have?

3.3.4 USE OF A DYNAMIC LOAD

The gain can be increased by the use of a dynamic load as shown in Figure 9(c) on p. 68. A current mirror is particularly effective with a long-tailed pair because it 'reflects' both the d.c. and the signal current of T1 into the collector lead of T2.

Consider the d.c. operation first. When there is no input signal the currents in T1 and T2 are equal. And since the currents in T3 and T4 of the current mirror are also equal, the current mirror 'reflects' just the required current into the collector lead of T2. So T4 and T2 are supplying and receiving the same d.c. current, with the result that there is none available for the load resistor R_L, giving zero voltage drop across it. In Figure 9(c) R_L is shown connected to a reference voltage of 4 V so that T2 and T4 normally have appropriate values of V_{CE}.

☐   Why might 0 V be an unsuitable reference voltage for R_L in this case?

■   Because if the d.c. level of the input is 0 V or more the collector of T2 will not be reverse biased and so T2 will not amplify properly.

Now when a small differential input voltage is applied, such that it raises the base voltage of T1, the current in T1 will increase whilst that of T2 will decrease. But T4 reflects the current in T1, so whilst the current in T2 decreases, the current in T4 increases. The difference between these two currents flows in R_L. The change in current in R_L is therefore twice what it was when R_{C2} was in the circuit. The same argument applies if the current in T1 decreases: the change in current in R_L is reversed but is still double what it would be with R_{C2} in the circuit. Hence, with the dynamic load, the gain of the circuit is now $g_m R_{L(total)}$, the same as for the common-emitter circuit.

In addition, however, $R_{L(total)}$ is now much higher than in Figure 9(a) because of the high output resistance of T4 (see Home Computing Exercise 4).

SAQ 14 (a) The circuit of Figure 9(c) is constructed from transistors that have the following parameters: For the p–n–p transistors: $\beta = 100$, $VA = 150$ V. For the n–p–n transistors: $\beta = 200$, $VA = 100$ V. For $I_S = 2$ mA find the value of R_L that will give a voltage gain of 500.

(b) Why does the use of a dynamic load, as in Figure 9(c), increase the CMRR?

3.3.5 THE EQUIVALENT CIRCUIT OF THE LONG-TAILED PAIR

All these results can be obtained a little more accurately by carrying out a nodal analysis of the equivalent circuit of the long-tailed pair. The equivalent circuits of each of the transistors have to replace the transistor symbols, as explained earlier.

SAQ 15 Draw the small-signal equivalent circuit of Figure 9(b). How many nodes are there?

Using the data of SAQ 14, estimate the differential input resistance of the circuit.

3.4 THE EMITTER-FOLLOWER

The basic emitter-follower circuit is shown in Figure 10(a). Figure 10(b) shows it coupled by capacitors to a voltage source (of internal resistance R_s) and to a load R_L. Notice that the signal output is taken across the resistor R_E in the emitter lead. The circuit gives a voltage gain of only about one ($A_v \approx 1$), but it has a large input resistance and a small output resistance so it can be used as a 'buffer'. Figure 10(c) shows a system diagram of a buffer with a 'black box' representing the emitter-follower. As explained in Block 1, because it has a high input resistance and a low output resistance the buffer can transfer almost the whole of V_s to the load even when the output resistance of the source is much bigger than the load resistance.

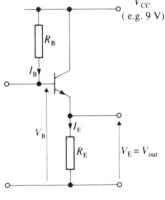

(a)

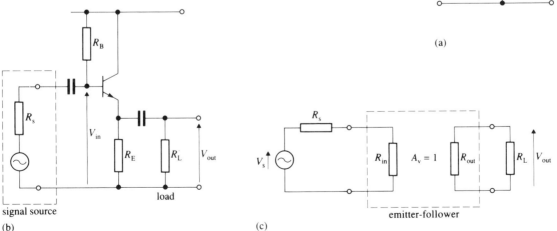

(b)

(c)

FIGURE 10 The emitter-follower: (a) the basic circuit; (b) an a.c., capacitor-coupled circuit; (c) a schematic diagram of the circuit

The operation of the circuit of Figure 10(b) can be understood as follows. Because the voltage across the emitter p–n junction does not vary much around 0.65 V as the input voltage varies, the output signal voltage 'follows' the input signal voltage—hence the name of the circuit. And because of the large current gain β of the transistor, the circuit can provide a much larger current into a load than a direct connection from source to load. Thus, for whatever signal current flows into the input terminal, almost β times as much is available in the output circuit.

3.4.1 THE D.C. CIRCUIT DESIGN

In the circuit of Figure 10(a) the base current needed to establish the operating point is supplied through R_B.

☐ Given that R_E in Figure 10(a) is 1 kΩ and that the typical value of β of the transistor used is 200, what is a suitable value of R_B to give a d.c. output voltage V_{OUT} of about half the supply voltage?

■ If the d.c. output voltage is to be 4 V it follows that I_E must be 4 mA, and the d.c. voltage of the base terminal is about 4.65 V because of the usual value of V_{BE}. Therefore the voltage drop across R_B is 4.35 V. Now I_B must be $I_C/\beta = 4$ mA/200 = 20 µA so $R_B = 4.35$ V/20 µA $\approx$ 220 kΩ.

The emitter resistance R_E can be thought of as a 'feedback' resistance because it has the effect of transferring the output voltage back into the input circuit. You would expect therefore that this circuit would not be so seriously affected by the spread of β as was the common-emitter amplifier. For example, a transistor with a higher value of β than 200 will produce a larger emitter current and therefore also a larger voltage drop across R_E. This means that the emitter and base voltages are higher and that the voltage drop across R_B is reduced. Thus I_B is reduced and the increase of I_E due to the higher value of β is somewhat diminished. As a consequence of this negative feedback effect, the circuit of Figure 10(a) can be satisfactory for the emitter-follower (though this kind of circuit with a single resistor R_B was not satisfactory for the common-emitter amplifier).

3.4.2 THE SMALL-SIGNAL EQUIVALENT CIRCUIT OF THE EMITTER-FOLLOWER

SAQ 16 Assuming that the input and output capacitors have negligible reactances at the low frequency of operation, draw the small-signal equivalent circuit of the emitter-follower of Figure 10(b).

The equivalent circuit of the emitter follower you have drawn for SAQ 16 is best analysed using your computer, since, unlike the example of the common-emitter amplifier, nodal analysis does not give rise to very convenient equations. However partly as an illustration of the paper-and-pencil method of analysis, and partly because the results obtained are not easy to arrive at using simpler methods, this circuit is analysed in the Appendix. You will not be expected to produce such an analysis yourself. The analysis shows that:

(i) the voltage gain is very nearly equal to 1;

$$A_v \approx 1 \tag{20}$$

(ii) if the source resistance is small (i.e. much less than r_i of the transistor equivalent circuit) the output resistance is r_e in parallel with R_E. And since normally $r_e \ll R_E$ this is approximately r_e. So

$$R_{out} \approx r_e (\approx 1/g_m) \tag{21}$$

☐ If I_E of the emitter-follower is 10 mA what is the circuit's output resistance when it is driven by a voltage source?

■ $g_m = KI_E$ so $g_m \approx 400$ mA/V. Therefore the output resistance $\approx 1/g_m = r_e = 2.5\Omega$.

(iii) if the source resistance R_s is *not* small compared with r_i the output resistance is approximately $(R_s + r_i)/\beta$ in parallel with R_E;

$$\frac{1}{R_{out}} = \frac{\beta}{R_s + r_i} + \frac{1}{R_E}. \tag{22}$$

(iv) the input resistance is given by

$$R_{in} = r_i + \frac{\beta}{g_L + g_E}. \tag{23}$$

So if $g_E \ll g_L$ (i.e. $R_L \ll R_E$) which is often the case,

$$R_{in} \approx r_i + \beta R_L$$

Example: In Figure 10(b), if $R_s = R_L = R_E = 1\ \mathrm{k\Omega}$ and the capacitors have negligible reactance and if β of the transistor is 200 and if I_E is set at 4 mA, what is the ratio of V_{out} to V_s?

With $I_E = 4\ \mathrm{mA}$ the transconductance g_m of the transistor is $160\ \mathrm{mA/V}$ and its input resistance r_i is $\beta/g_m = 1250\ \Omega$.

(i) From equation (23), remembering that $R_E = R_L = 1\ \mathrm{k\Omega}$, the input resistance of the emitter follower as a whole becomes

$$R_{in} = 1250\ \Omega + \frac{200}{0.001\ \mathrm{S} + 0.001\ \mathrm{S}} \approx 101\ \mathrm{k\Omega}.$$

(Note that r_i is relatively very small.)

(ii) By equation (22) the output resistance is given by

$$\frac{1}{R_{out}} = \frac{200}{1\ \mathrm{k\Omega} + 1.25\ \mathrm{k\Omega}} + \frac{1}{1\ \mathrm{k\Omega}} = 0.089\ \mathrm{S} + 0.001\ \mathrm{S} = 0.09\ \mathrm{S}$$

or $R_{out} = 11\ \Omega.$

(Note that the presence of a source resistance of 1 kΩ has caused R_{out} to be significantly bigger than r_e.)

Putting these figures for R_{in} and R_{out} into the system diagram of Figure 10(c) gives

$$\frac{V_{out}}{V_s} = \frac{101\ \mathrm{k\Omega}}{102\ \mathrm{k\Omega}} \times 1 \times \frac{1000\ \Omega}{1011\ \Omega} = 0.98.$$

This compares with $V_{out}/V_s = 0.5$ if the emitter follower were not there.

SAQ 17 In Figure 10(b) the emitter-follower circuit is operating as in the above example, but the internal resistance of the source is 10 kΩ and the external load resistance is 50 Ω. What is V_{out}/V_s?

A d.c. coupled version of the emitter-follower is shown in Figure 11. Like the d.c. coupled version of the common-emitter amplifier it is intended for inclusion in an amplifier which has overall negative feedback so that the d.c. operating point can be automatically adjusted (see Section 4). The resistor R_E is replaced by a current source so that almost all the signal output current flows in R_L instead of being shared between R_L and R_E. R_B is included to give a V_{BE} of 0.65 V for the design value of I_{IN} as explained in the next section.

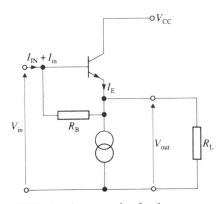

FIGURE 11 An example of a d.c. coupled emitter-follower circuit

4 A SIMPLE OPERATIONAL AMPLIFIER

4.1 THE BASIC CIRCUIT DESIGN

The d.c. coupled circuits described in Section 3 can be combined to form an operational amplifier as shown schematically in Figure 12(a). Figure 12(b) shows a more detailed circuit diagram. The first stage is the *long-tailed pair* of Section 3.3 with a *current source* in its 'tail'. This gives the amplifier a differential input and a fairly large CMRR. The long-tailed pair is d.c. coupled to the *common-emitter amplifier* which is made from a p–n–p transistor instead of an n–p–n one (so it 'hangs' from the positive supply line instead of 'standing' on the negative supply line). This circuit gives a large voltage gain, of more than 1000, because it has a dynamic load consisting of the high input resistance of an *emitter-follower*. This emitter follower provides the low output resistance needed to drive subsequent circuits satisfactorily.

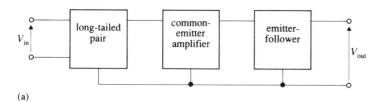

(a)

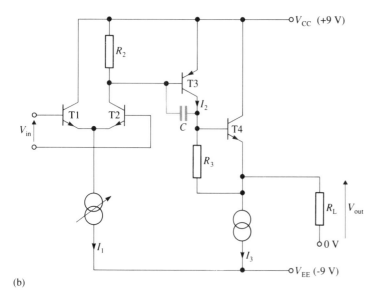

(b)

FIGURE 12 A simple op amp comprising a long-tailed pair, a common-emitter stage and an emitter-follower at the output: (a) a schematic diagram; (b) the circuit diagram

4.1.1 THE D.C. OPERATING POINTS OF THE TRANSISTORS

The final output voltage of an op amp is designed to be 0 V when there is no input signal. Because all the stages are d.c. coupled it is possible to make only a single d.c. adjustment at the input to produce the required zero output voltage. At the same time this adjustment also brings the intermediate transistors to their intended operating points despite their variations in β etc. This is called the **zero off-set adjustment** and it usually consists of adjusting I_1 the current supplied by the tail of the long-tailed pair as indicated by the arrow through the current-source symbol in Figure 12(b).

The way in which the current source I_1 can be adjusted is illustrated in Figure 13(a). This shows the input end of the op amp with the current source I_1 drawn as a current mirror. I_1 can evidently be adjusted by varying R_1 without interfering with the amplifying part of the circuit. Usually only small adjustments are needed, so R_1 might be a large fixed resistance in series with a small variable one.

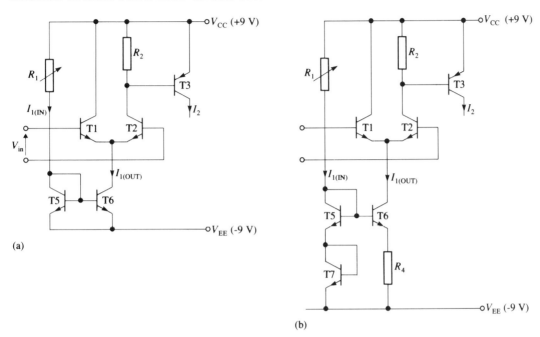

FIGURE 13 The input stage of the op amp: (a) with a normal current mirror as the current source; (b) the same but with a current source whose temperature dependence balances that of T3

The d.c. operating currents of the three stages can be chosen to match the application. Typically d.c. currents that increase by a factor of 10 at each stage are chosen. So if the first current source were designed to provide a current of 100 μA, then I_2 would be designed to be 1 mA and the second current source would be designed to provide $I_3 = 10$ mA.

For these d.c. operating currents the resistances of the two resistors R_2 and R_3 in Figure 12(b) can be calculated as follows. The emitter junction of T3 requires a forward bias of about 0.65 V, so, since the collector current of T2 is 50 μA and the base current of T3 is likely to be about 5 μA, R_2 must be about 0.65 V/45 μA = 14.5 kΩ. Similarly, since the collector current of T3 is to be 1 mA and the base current of T4 is likely to be about 50 μA, R_3 must be about 0.65 V/0.95 mA = 685 Ω in order to forward bias the emitter junction of T4 to the correct level.

This completes the d.c. design. If the amplifier has one of the negative feedback circuits described in previous blocks connected round it, any small errors in the offset adjustment due to temperature variations or to drift in parameter values will normally be reduced to negligible proportions.

4.1.2 ESTIMATING THE PERFORMANCE OF THE CIRCUIT

Now when a signal is applied to the differential input an amplified signal current appears at the emitter of T4 and, since the d.c. current source producing I_3 has a high resistance, almost all the signal current flows in the load resistor R_L giving the output voltage.

The voltage gain of the circuit is the product of the voltage gains of the three stages. The emitter-follower gives a voltage gain of 1, so the large gain of the circuit as a whole comes from the other two stages (see Exercises 5, 6 and 8 in the Home Computing Booklet for this block).

SAQ 18	If the operating currents in the circuit of Figure 12(b) are $I_1 = 100\ \mu A$, $I_2 = 1\ mA$ and $I_3 = 10\ mA$, and if the β and Early voltage of each transistor are 200 and 100 V, respectively, estimate (a) the differential input resistance of the amplifier; (b) the output resistance of the amplifier as seen by the load; (c) the open-circuit voltage gain of the amplifier (i.e. with $R_L = \infty$) by first calculating the gain of the long-tailed pair, and then that of the common-emitter stage.

SAQ 19	In carrying out the calculations for SAQ 18 you will have noticed that the voltage gain provided by the long-tailed pair is rather small. Why is this and how do you think it might be increased?

The capacitor C, which appears grey in Figure 12(b), across the collector junction of T3, may be included to provide frequency compensation. That is, it produces a single lag open-loop frequency response of the type described in Block 2, Part 3. This will ensure stability when frequency-independent negative feedback is applied around the complete amplifier. Placing the capacitor between collector and base of T2, where most of the gain is generated, is the optimum position for it since its capacitance can be relatively small. If it were placed across the output, for example, it would have to have a much larger capacitance to have a comparable effect (see Block 8 for further discussion).

4.2 ADDITIONAL DESIGN CONSIDERATIONS

4.2.1 RAIL REJECTION

An important feature of a high-gain amplifier is its ability to reject unwanted input signals; in particular any residual voltage waveforms which might be induced across the positive or negative supply lines, such as mains hum (either 50 Hz or 100 Hz or both) or the transient voltages caused by mains switches being operated elsewhere. Rejecting such signals is called **rail rejection**. It is achieved in this amplifier as follows:

(i) Signal voltages between the negative supply line and 0 V are blocked by the current generators. The current mirror which forms the current source $I_{1\,(OUT)}$ is shown in Figure 13(a). Evidently any negative-rail signal voltages affect the two transistors of the current mirror equally and so do not affect the forward bias of T6. The d.c. current it supplies is therefore hardly affected.

(ii) Similarly, signal voltages on the positive supply line are *almost* prevented from entering the amplifier because they affect the base and emitter of T3 *almost* equally. R_2 and the output resistance of T2 form a potential divider, and since R_2 is much the smaller of the two resistances the rail signal voltage at the base of T3 is almost the same as that at the emitter terminal, so only a small fraction of any change of voltage on the positive rail appears across the input T3. Note that this is another benefit of including R_2 in the circuit. (See Exercise 5 in the Computing Booklet.)

4.2.2 TEMPERATURE COMPENSATION

Because of the high gain of the circuit any variations of V_{BE} in T1, T2 or T3 due to temperature changes can affect the output quite seriously, unless they are balanced out. The two transistors T1 and T2 in the long-tailed pair are well balanced, so that any temperature change there produces a common-mode voltage which is already well rejected as explained in Section 3.3. However a rise in temperature of T3 will cause an increase in I_2, and a corresponding large change in output voltage, unless a method of compensation is introduced. One way of doing this is to ensure that the current source $I_{1\,(OUT)}$, feeding the long-tailed pair, produces a current that *decreases* with increasing temperature. Figure 13(b) shows how this can be done.

Transistor T7 and resistor R_4 are added to the current mirror. R_4 is chosen to drop the same voltage as the emitter junctions of T5, T6 and T7, namely about 0.65 V, at a particular temperature, so that $I_{1\,(IN)} = I_{1\,(OUT)}$ at that temperature (e.g. if $I_{1\,(OUT)}$ is intended to be 100 μA then $R_4 = 6.5$ kΩ). As the temperature rises the voltage drop across the three emitter junctions decreases by about 2 mV per degree, but the voltage drop across R_4 does not change significantly. So the forward bias of the emitter-base junction of T6 decreases as the temperature rises; hence $I_{1\,(OUT)}$ decreases correspondingly, as required. Because the *causes* of temperature drift in opposite directions of $I_{1\,(OUT)}$ and I_2 are the same they balance quite accurately, and the amplifier performance is not much affected by temperature.

4.2.3 SOME COMMENTS ON THE 741 OP AMP

The type 741 operational amplifier referred to in Block 2 has the same basic three stages as that in Figure 12: namely a long-tailed pair, followed by a high gain common-emitter amplifier, followed by a low output impedance stage. The long-tailed pair has a dynamic load for increased gain and improved CMRR, with an emitter-follower on its output to couple it to the common-emitter stage. These two high-gain stages provide an overall gain of around a million. The output stage is also driven by an emitter-follower to keep the output resistance low. The circuit includes many additional subtleties which need not concern us here.

5 ANALOGUE CIRCUITS USING MOSFETs

As you might expect, since their equivalent circuits are similar, you can construct similar circuits to those described in Sections 3 and 4 using MOSFETs instead of bipolar transistors. That is you can make *common-source amplifiers*, *source-followers* and *long-tailed pairs* using MOSFETs with similar properties to the corresponding biplolar circuits. The main differences are (i) no gate current is required; (ii) precisely matched pairs, with the same threshold voltages and transconductances, as required for long-tailed pairs and dynamic loads, are more difficult to produce, and (iii) the voltage gain available per stage in integrated circuits is less.

☐   Under what operating conditions can the gain available from a MOSFET common-source amplifier be maximised?

■   By operating at a low drain current. Remember that the maximum available gain is proportional to $1/\sqrt{I_D}$.

For these reasons, especially for the increased gain they offer, bipolars are usually preferred for analogue circuits. The main exception is as an input to an op amp. The fact that no d.c. gate current is required implies that a very large low-frequency input resistance can be obtained with MOSFETs.

MOSFETs come into their own in digital circuits, so MOSFET circuits are discussed more fully in Part 3. However the simplicity of MOSFET circuits has much to commend it, even in analogue circuits.

The simplicity emerges most strongly when 'complementary pairs' of transistors are used (i.e. an n-channel and a p-channel MOSFET with equal transconductances and equal but opposite threshold voltages). The family of circuits based on complementary pairs of MOSFETs is known as CMOS for short. This form of circuitry has achieved pre-eminence mainly in digital circuits, but analogue CMOS is also very effective as illustrated below.

Figures 14(a) and (b) show two possible designs of CMOS amplifiers. In Figure 14(a) the drain 'resistor' for transistor T1 is a p-channel depletion-mode transistor, T2. With its gate connected to its source the current I_{DSS} flows through it, giving a higher small-signal resistance than a resistor would for the same d.c. current. So T2 is a kind of dynamic load.

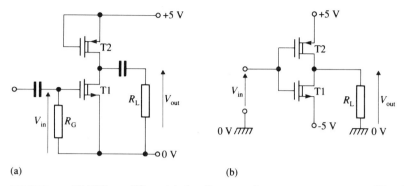

(a)   (b)

FIGURE 14 CMOS amplifiers: (a) the diagram of a common-source amplifier in which a p-channel MOSFET is used as a dynamic load; (b) the circuit diagram of a push-pull CMOS amplifier

☐   Why must T1 and T2 in Figure 14(a) be depletion-mode MOSFETs?

■   Because it is necessary for current to flow through them even when, in each case, $V_{GS} = 0$.

The gate voltage of T1 is held at 0 V by the resistor R_G which can be very large because there is virtually zero gate current. Evidently the MOSFETs must be fairly closely matched so that they pass about the same I_{DSS}.

Figure 14(b) shows a d.c. coupled amplifier, which is as simple an arrangement as it is possible to produce! Both transistors contribute to the gain of the circuit since the signal input is applied to both gates. When correctly adjusted the currents in the two transistors are the same so that there is no current in the load and the output voltage is zero. Then any small additional voltage applied to the input will cause the drain current in one transistor to increase as the current in the other decreases. The current in the load is the difference between these two currents. Such a balanced circuit is called a **push-pull** amplifier.

When built into a high-gain amplifier the d.c. levels can be maintained at the correct operating point by overall negative feedback. In this circuit there is a second active transconductance, so that the change in output current per unit change of input voltage is doubled. Since each transistor contributes to the gain,

$$\text{the voltage gain } A_v = -(g_{m1} + g_{m2})R_{L\,(\text{total})} \qquad (24)$$

where as usual $R_{L\,(\text{total})}$ is the net load resistance. That is

$$\frac{1}{R_{L\,(\text{total})}} = g_{o1} + g_{o2} + g_L.$$

As explained in Section 2.2.2 the gain can become quite large if the transistor is operated close to its threshold voltage (i.e. when I_D is small). However the larger the gain achieved in this way, the less predictable it is because of the difficulty of matching threshold voltages.

6 CONCLUSION

This text has only considered some basic analogue circuits; there are of course many other useful and effective transistor circuits that can be designed. The aim of the text has simply been to introduce some commonly used circuits and to show how they can be designed to have certain specified properties. For a fuller study of basic transistor circuits, two books can be recommended. A recent introductory book is *Transistor Circuit Techniques: Discrete and Integrated* by G. J. Ritchie, published by Van Nostrand Reinhold. A classic comprehensive book is *The Art of Electronics* by Horowitz and Hill, first published in 1980 by The Cambridge University Press (a second edition of this book was published in 1989).

7 SUMMARY

(Note: The equations in this summary are the only formulae you will be expected to remember in examinations.)

1   The *d.c. operating currents* and *voltages* of transistors must be established before their *small-signal equivalent-circuit parameter values* can be determined, and small-signal analysis can begin.

2   Transistors cannot be made with perfect uniformity. The parameter *spreads* which are most marked are the current gain β in bipolar transistors, and the threshold voltage V_T and the d.c. current for a given gate voltage in MOSFETs. The various designs of the basic circuits differ in the extent to which their d.c. operating points and a.c. performance are affected by these parameter spreads. Feedback around high-gain amplifiers containing d.c. coupled circuits provides one of the best ways of controlling not only the operating points and small-signal performance but also the temperature dependence, distortion, etc.

3   The basic *low-frequency, small-signal equivalent circuit of a transistor* is as shown again in Figure 15.

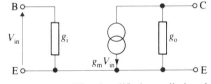

FIGURE 15   The simplified, small-signal, low-frequency equivalent circuit of a transistor

(a)   For a bipolar transistor

$$g_m = K I_C$$

where $K = 40$ V^{-1} at room temperature, $g_i = 1/r_i \approx g_m/\beta$, or to be more precise, g_m/h_{fe}, $g_o = 1/r_o \approx I_C/(VA + V_{CE})$ where VA is the Early voltage. β is the d.c. common-emitter current gain, and h_{fe} is the small-signal common-emitter current gain. At high frequencies h_{fe} decreases with frequency (see Block 8) I_C is the collector d.c. operating current and V_{CE} is the operating collector-emitter voltage.

Typically, $\beta \approx h_{fe}$ is in the range 100 to 500.

(b)   For a MOSFET

$$g_i = 0.$$

In the *saturation region of operation* (not to be confused with the saturation region of bipolar transistors)

$$g_m = \beta(V_{GS} - V_T)(1 + \lambda V_{DS})$$

$$g_o = \frac{\lambda I_D}{1 + \lambda V_{DS}} \approx \lambda I_D$$

where β is the gain factor (not to be confused with the current gain β of bipolar transistors), V_T is the threshold voltage, λ is the channel length modulation factor, I_D is the operating drain current, and V_{GS} and V_{DS} are the operating gate and drain voltages.

Typically, when $I_D = 1$ mA, $g_m = 1$ mA/V and $r_o = 10$ kΩ.

In the *linear region*

$$g_m = \beta V_{DS}$$
$$g_o = \beta(V_{GS} - V_T - V_{DS}).$$

4   The voltage gain of a common-emitter or common-source amplifier is $-g_m R_{L\,(total)}$, where $R_{L\,(total)}$ is the resistance of the parallel combination of load conductance g_L, the transistor's output conductance, g_o, and the conductance g_R of the resistor or other circuit element in the collector or drain lead.

Thus

$$A_v = -\frac{g_m}{g_o + g_R + g_L}.$$

This has a maximum possible value of $-g_m/g_o$ which can be several thousand for a bipolar transistor, but may be only a few tens for a MOSFET. In a MOSFET amplifier A_v is proportional to $1/\sqrt{I_D}$. The voltage gain of a bipolar amplifier can be greatly increased by the use of a dynamic load such as a current mirror or an emitter-follower.

5   The differential voltage gain of a long-tailed pair $\approx \pm g_m R_{L\,(total)}/2$.

Made from bipolars, the circuit has a low-frequency differential input resistance of $2\beta/g_m$. Made from MOSFETs its low-frequency differential input resistance is virtually infinite.

The voltage gain of a long-tailed pair can be increased by the use of a dynamic load in the form of a current mirror. This increases $R_{L\,(total)}$ and gives a differential gain of $g_m R_{L\,(total)}$.

The common-mode gain of a long-tailed pair is $\approx R_C/2R_E$ where R_E is the resistance of the long tail. The differential voltage gain $\approx -g_m R_C/2$, so the common-mode rejection ratio, which is the ratio of these two gains is (CMRR) $\approx -g_m R_E$. The CMRR can be increased by replacing resistor R_E with a current source and by using a dynamic load.

6   The voltage gain of an emitter-follower is approximately 1. Its input resistance is usually approximately $\beta R_{L\,(total)}$, where $R_{L\,(total)}$ is the parallel combination of the resistance of the emitter resistor and the external load resistance. Its output resistance is usually about $1/g_m$ unless the source resistance is large, in which case the output resistance is greater than $1/g_m$.

7   The great attraction of CMOS circuits is their simplicity and their high input resistance. Their voltage gain is, however, normally relatively small compared with bipolar circuits.

8   The voltage gain of an op amp is the product of the voltage gains of the various stages of the amplifier. These stages include a long-tailed pair to provide a differential, high-resistance input; a common-emitter high-gain stage and finally a low-output-resistance output stage such as an emitter follower. Its design includes maximizing common-mode rejection and rail rejection and minimizing temperature dependence. The circuit is d.c. coupled throughout so that overall feedback can be used to stabilize the d.c. operating point and reduce temperature dependence as well as to control the amplifier's small-signal properties.

ANSWERS TO SELF-ASSESSMENT QUESTIONS

SAQ 1

(a) The 'd.c. operating point' of a nonlinear resistor is simply the voltage across it, and/or the current flowing through it. The d.c. conductance of the device is then the d.c. current flowing through it divided by the d.c. voltage across it. The small-signal conductance is *the slope* of the graph of d.c. current versus d.c. voltage measured at the operating point.

From the graph of Figure 2 it is easy to see that both the d.c. conductance and the small-signal conductance change if the operating point is changed.

(b) The d.c. conductance when $V = 6$ V is about 4.6 mA/6 V = 0.77 mS. The slope of the graph (i.e. the slope of the tangent) at the operating point is about 4.6 mA/4 V = 1.15 mS.

SAQ 2

Using the numerical data and the theoretical equations:

From equation (2), $g_m = 40$ V$^{-1} \times 1$ mA = 40 mA/V.

From equation (3) and the fact that $\beta = 100$, $g_i = g_m/\beta = 400$ μS.

From equation (5) and the fact that $VA = 150$ V,

$$g_o = \frac{I_C}{VA + V_{CE}} = \frac{1 \text{ mA}}{153 \text{ V}} = 6.5 \text{ μS}.$$

SAQ 3

The equivalent-circuit values are as follows:

from equation, (2) $g_m = KI_C = 40$ V$^{-1} \times 0.1$ mA = 4 mA/V;
from equation, (3) $g_i = g_m/\beta = 4$ mA/V $\div 250 = 1.6 \times 10^{-5}$ S or $r_i = 62.5$ kΩ;

from equation (5), $g_o = I_C/(VA + V_{CE}) = 0.1$ mA/105 V = 0.95 μS, or $r_o = 1.05$ MΩ.

SAQ 4

β is the ratio g_m/g_i, or $g_m r_i$. So in this case $\beta = 20$ mA/V $\times 8000$ Ω = 160. The value of the Early voltage can be obtained directly from equation (5). Substituting the values given gives

$$\frac{1}{300\,000 \text{ Ω}} = \frac{0.5 \text{ mA}}{VA + 5 \text{ V}}, \text{ so } VA = 145 \text{ V}.$$

The equivalent-circuit values can now be found as in SAQ 3:

$g_m = 40$ V$^{-1} \times 0.2$ mA = 8 mA/V;

$g_i = 8$ mA/V $\div 160 = 50$ μS, or $r_i = 20$ kΩ;

$g_o = I_C/(VA + V_{CE}) = 0.2$ mA/155 V = 1.3 μS or $r_o = 0.78$ MΩ.

SAQ 5

Again accurate results cannot be expected here.

(a) g_o at the specified operating point, is the slope of the $+0.5$ V curve in Figure 5(a) at $V_{DS} = 5$ V. This is about 0.4 mA/10 V. So $g_o \approx 40$ μS, or $r_o = 25$ kΩ.

(b) g_m is the slope of the 5 V curve in Figure 5(b) at $V_{GS} = 0.5$ V. This is about 4 mA/2.6 V. So $g_m \approx 1.5$ mA/V.

The threshold voltage V_T is where the curves meet the voltage axis on Figure 5(b), since this is the gate voltage at which drain current starts to flow. It is about -1.6 V.

SAQ 6

(a) From equation (6), for the linear region:

$$I_D = \beta\{(V_{GS} - V_T) - V_{DS}/2\}V_{DS}.$$

Differentiating this equation with respect to V_{GS}, with β, V_T and V_{DS} constant, gives

$$g_m = dI_D/dV_{GS} = \beta V_{DS}.$$

Similarly, differentiating equation (6) with respect to V_{DS}, with V_{GS} constant gives

$$g_o = dI_D/dV_{DS} = \beta\{(V_{GS} - V_T) - V_{DS}\}.$$

These are the required equations.

(b) At $I_D = 1.5$ mA and $V_{DS} = 1$ V you can see from Figure 5(a) that $V_{GS} = 1$ V. You have already established that $V_T = -1.6$ V and that $\beta = 0.72$ mA V^{-2}, so

$$g_m = 0.72 \text{ mA V}^{-2} \times 1 \text{ V} = 0.72 \text{ mA/V},$$

and

$$g_o = 0.72 \text{ mA V}^{-2} \times (1 \text{ V} + 1.6 \text{ V} - 1 \text{ V})$$
$$= 0.72 \text{ mA V}^{-2} \times 1.6 \text{ V} = 1.15 \text{ mS},$$

or $r_o \approx 870$ Ω.

SAQ 7

The operating point is $V_{CE} = 4.5$ V and $I_C = 2$ mA so $g_m = 80$ mA/V.

The input conductance $g_i = g_m/\beta = 80$ mA/V $\div 100 = 800$ μS (or $r_i = 1250$ Ω).

The Early voltage is 100 V, so $g_o = I_C/(VA + V_{CE}) = 2$ mA/104.5 V = 19 μS or $r_o \approx 52$ kΩ.

SAQ 8

The small-signal equivalent circuit is shown in Figure 16. It is the same as Figure 6(d) except that the source conductance g_s is added.

V_{in} is less than 5 mV because of the generator's internal resistance: that is

$$V_{in} = \frac{V_s \times R_{in}}{R_{in} + R_s}.$$

R_{in} is given by equation (14), where g_B is so small that it can be ignored in comparison with g_i, so $R_{in} \approx r_i = 1.25$ kΩ. Therefore

$$V_{in} = \frac{5 \text{ mV} \times 1.25 \text{ kΩ}}{1.25 \text{ kΩ} + 1 \text{ kΩ}} = 2.8 \text{ mV}.$$

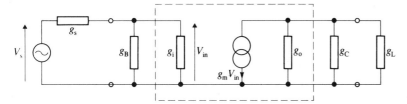

FIGURE 16 The equivalent circuit of a common emitter amplifier

The net load resistance $R_{L\,(total)}$ is the parallel combination of the resistances of R_C, r_o and R_L. Their conductances are respectively 455 μS, 19 μS and 125 μS. The operating current is 2 mA so $g_m = 80$ mA/V, so, from equation (12)

$$A_v = -\frac{0.08}{(455 + 19 + 125) \times 10^{-6}} = -134.$$

Therefore $V_{out} = -2.8$ mV $\times 134 = -0.38$ V.

SAQ 9

In Figure 6(a), when $\beta = 100$, $I_C = 2$ mA and $V_{CE} = 4.4$ V. When however $\beta = 200$, $I_C = 4$ mA and the voltage drop across R_C is 8.8 V. So $V_{CE} = 9$ V $- 8.8$ V $= 0.2$ V. The transistor is therefore driven into the very curved part of the d.c. output characteristics which is called the *saturation region of operation*. See Figure 4(d) at $V_{CE} = 0.2$ V. The output waveform will be very distorted due to clipping. If $\beta > 200$ the transistor is driven even further into saturation and the clipping is even more severe. (The saturation region of operation is described and explained in some detail in Part 3 of this block.)

SAQ 10

The fact that the output current from the transistor flows through R_E and produces a voltage at the input means that the feedback is current-derived. The fact that it subtracts directly from the input signal voltage means that the feedback is series connected to the input. It therefore increases the input resistance of the circuit.

SAQ 11

The technique is to connect the capacitor in parallel with only part of R_E. Thus if R_E is split to form two resistors R_{E1} and R_{E2} and only the lower one is bypassed by the capacitor, as in Figure 17, the gain of the circuit will be $R_C/(R_{E1} + r_e)$. So, to obtain a gain of -22 when $R_C = 2.2$ kΩ, $R_{E1} + r_e$ must be 100 Ω. But $r_e = 1/g_m = 12.5$ Ω, so $R_{E1} = 87.5$ Ω, leaving $R_{E2} \approx 413$ Ω.

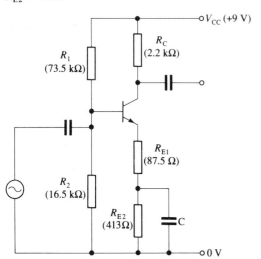

FIGURE 17 A common emitter amplifier with a voltage gain of -22

SAQ 12

The d.c. supplies are $+9$ V and -9 V so the voltage drop across R_1 is about 17.35 V.

The current through R_1 is to be 1 mA, so $R_1 = 17.35$ V/1 mA $= 17.35$ kΩ.

Since the collector current of T1 is to be 1 mA, g_m of T1 is 40 mA/V.

With zero input signal and zero d.c. output voltage, V_{CE} of T1 and T3 is 9 V, so the output conductances of T1 and T3 are

$$g_o = \frac{I_C}{VA + V_{CE}} = \frac{1\text{ mA}}{159\text{ V}} = 6.3\ \mu S.$$

(a) From equation (16), when $R_L = 0.5$ MΩ

$$\text{the voltage gain } A_v = -\frac{40\text{ mA/V}}{(6.3 + 6.3 + 2.0) \times 10^{-6}\text{ S}}$$

$$\approx -2740.$$

(b) When $R_L = 5000$ Ω

$$A_v = -\frac{40\text{ mA/V}}{(6.3 + 6.3 + 200) \times 10^{-6}\text{ S}} \approx -188.$$

Evidently the load resistance has a great effect on the gain.

SAQ 13

(a) The current in each transistor will be 100 μA so $g_m = 4$ mA/V. The voltage drop across R_E is about 8.3 V, so the resistance of R_E has to be 8.3 V $\div$ 200 μA ≈ 42 kΩ. Therefore the CMRR $= -4$ mA/V $\times 42$ kΩ $= -168$.

(b) Figure 18 shows the circuit you should have drawn.

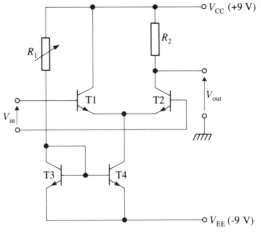

FIGURE 18

To obtain a larger CMRR the output resistance of the current source must be much larger than 42 kΩ. Referring to Figure 18, the output resistance r_{o4} of T4 must be such that $g_{m2}r_{o4} = 3000$ (i.e. r_{o4} replaces R_E in equation (19)). But g_{m2}, the transconductance of T2, $= 4$ mA/V, so $r_{o4} = 3000/(4\text{ mA/V}) = 750$ kΩ.

The output resistance of a transistor is given by equation (5). Thus

$$r_o = 1/g_o = (VA + V_{CE})/I_C.$$

In this case, where $V_{CE} \approx 8$ V, and $I_C = I_S$,

$$750 \text{ k}\Omega = (VA + 8 \text{ V})/200 \text{ μA}.$$

So, for transistor T4, VA should be 142 V.

SAQ 14

(a) The current source provides a current of 2 mA. The operating collector current of all four transistors is therefore 1 mA; so their transconductances are 40 mA/V. To calculate the voltage gain the output conductances g_{o2} and g_{o4} of T2 and T4 must first be calculated. T4 is a p–n–p transistor and $V_{CE} = 9 \text{ V} - 4 \text{ V} = 5 \text{ V}$, so $g_{o4} = I_C/(VA + V_{CE}) = 1 \text{ mA}/155 \text{ V} = 6.5$ μS.

T2 is an n–p–n transistor, so $g_{o2} = 1 \text{ mA}/104 \text{ V} = 9.6$ μS.

The voltage gain $= g_m/(g_{o2} + g_{o4} + g_L)$, whose value in this case is to be 500. So the equation to be solved is

$$\frac{40 \text{ mA/V}}{(9.6 + 6.5) \times 10^{-6} \text{ S} + g_L} = 500$$

or

$$(9.6 + 6.5) \times 10^{-6} \text{ S} + g_L = \frac{40 \text{ mA/V}}{500} = 80 \times 10^{-6} \text{ S}$$

so

$$g_L = 63.9 \text{ μS}$$

giving

$$R_L = 15.6 \text{ k}\Omega.$$

(b) The dynamic load increases the *differential* gain, as already explained, but it does not increase the *common-mode* gain, for the following reason. Assume that the circuit is properly set up so that there is zero current in the load. Consider what happens when a positive-going common-mode voltage is applied to the input terminals. This will produce the same small increase in current in the input transistors T1 and T2, as a result of the increase in voltage drop across the current source. The increase in the currents in T1 and T2 means that the currents in the current mirror also increase equally by the same amount. So the currents in T2 and T4 in Figure 9(c) increase by equal amounts, and the current flowing in the external load R_L remains at zero. So to a first approximation the common-mode gain is reduced to zero by the dynamic load. Imbalances in the transistors will give rise to *some* common-mode gain, but the CMRR will certainly be increased.

SAQ 15

The required equivalent circuit is shown in Figure 19. There are evidently 4 nodes in addition to the earth node.

It is clear from the equivalent circuit diagram that the resistance between the two input terminals is $2r_i$ or $2/g_i$. From the data for n–p–n transistors given in SAQ 14, $I_S = 2$ mA so the transconductances of T1 and T2 are 40 mA/V, and their βs are 200. So r_i for each input transistor is $\beta/g_m = 5$ kΩ. The input resistance R_{in} of the long-tailed pair is therefore 10 kΩ.

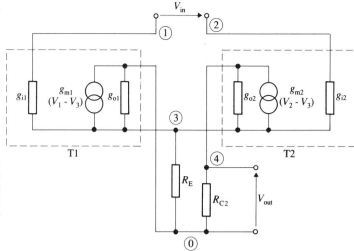

FIGURE 19 The small-signal equivalent circuit of a long-tailed pair

SAQ 16

See Figure A2 in the Appendix.

SAQ 17

Since $R_L \ll R_E$:

$$R_{in} \approx r_i + \beta R_L = 1250 \ \Omega + 200 \times 50 \ \Omega = 11\,250 \ \Omega.$$

R_{out} is given by equation (22), namely,

$$\frac{1}{R_{out}} = \frac{\beta}{R_s + r_i} + \frac{1}{R_E}$$

so

$$\frac{1}{R_{out}} = \frac{200}{10 \text{ k}\Omega + 1.25 \text{ k}\Omega} + \frac{1}{1 \text{ k}\Omega}$$

$$= 0.0178 \text{ S} + 0.001 \text{ S} = 0.0188 \text{ S}.$$

So $R_{out} = 53 \ \Omega$.

Therefore, referring to Figure 10(c),

$$\frac{V_{out}}{V_s} = \frac{11.25 \text{ k}\Omega}{21.25 \text{ k}\Omega} \times 1 \times \frac{50 \ \Omega}{103 \ \Omega} = 0.26.$$

Despite the buffer only a quarter of V_s appears across the load. But without the buffer $V_{out}/V_s = 50/10\,050 \approx 0.005$!

SAQ 18

(a) The differential input resistance of a long-tailed pair is the input resistances of the two input transistors in series, namely $2 \times r_i$. With 50 μA operating currents in the transistors, $g_m = 2$ mA/V. Therefore with $\beta = 200$, $2r_i = 2\beta/g_m = 200$ kΩ.

(b) The output resistance of the circuit is the output resistance of the emitter-follower, T4 which is driven by the output of T3. The output resistance of T3 is r_o of T3.

Now r_o of T3 can be calculated from its Early voltage $VA = 100$ V.

Thus $r_o = (VA + V_{CE})/I_C \approx 109 \text{ V}/1 \text{ mA} = 109$ kΩ.

The output resistance of an emitter-follower driven by a high input resistance is given by equation (22). Here R_E is in effect very large so it can be ignored, and r_i of T4 is much less than 109 kΩ, so

$$R_{out} \approx (r_o \text{ of T3})/\beta \approx 540 \ \Omega.$$

(c) The gain of the long-tailed pair is given by $g_m R_{L\,(total)}/2$.

Since I_C of T2 is 50 μA, $g_{m2} = 2$ mA/V.

$R_{L\,(total)}$ is R_2 in parallel with the input resistance of T3 and the output resistance of T2, thus

$$\frac{1}{R_{L\,(total)}} = \frac{1}{R_2} + \frac{g_{m3}}{\beta} + \frac{I_{C2}}{VA + V_{CE2}}.$$

Since the collector current of T3 is 1 mA, $g_{m3} = 40$ mA/V.

$V_{CE2} \approx 8$ V if the d.c. level of the two input terminals is 0 V, so

$$\frac{1}{R_{L\,(total)}} = \frac{1}{14.5\ k\Omega} + \frac{40\ mA/V}{200} + \frac{50\ \mu A}{108\ V}$$

$$= (69 + 200 + 0.5) \times 10^{-6}\ \mu S$$

$$= 270\ \mu S$$

so $R_{L\,(total)} \approx 3700\ \Omega$.

Therefore the voltage gain of the long-tailed pair is $2\ mA/V \times 3700\ \Omega \div 2 = 3.7$.

The voltage gain of the common-emitter stage T3 is $g_m R_{L\,(total)}$.

The load of T3 is the input resistance of the emitter follower which, on open circuit is very high and can be ignored. (R_3 is in series with the open circuit load and with the current source so does not affect the load on T3.) So the effective load of T3 is simply its own output resistance, namely (by equation (5)), 108 V/1 mA = 108 kΩ.

The transconductance of T3 is 40 mA/V,

so voltage gain of T3 = 40 mA/V $\times$ 108 kΩ = 4320.

The gain of the emitter-follower is 1.

Therefore the overall open circuit gain

$$A_v = 3.7 \times 4320 \times 1 \approx 16\,000.$$

SAQ 19

The voltage gain of the long-tailed pair is small because its total load resistance R_2 is small. R_2 is chosen so that the d.c. voltage drop across it is only 0.65 V. As explained in Section 3.3 the voltage gain of a long-tailed pair can be greatly increased by replacing the load resistance with a current mirror as dynamic load. (Additional changes would also be needed to obtain the right d.c. operating points for transistors T2 and T3 but these need not concern us here.) If gains of 1000 or so can be obtained from both the long-tailed pair and the common emitter stage an overall gain of a million results.

APPENDIX: AN ANALYSIS OF AN EMITTER-FOLLOWER

Figure A1 is the circuit diagram of a general emitter-follower with signal source and load. Figure A2 shows its equivalent circuit with the nodes numbered. The following analysis refers to Figure A2.

You will not be assessed on this material.

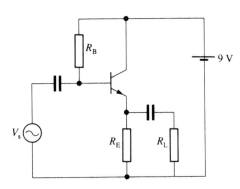

FIGURE A1 The circuit of an emitter-follower

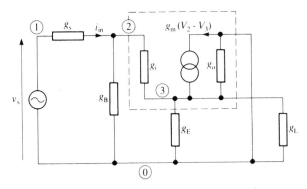

FIGURE A2 The small-signal equivalent circuit of an emitter-follower

(a) *Input conductance* is given by i_{in}/v_{in}. The nodal equations are as follows:

at node 2,

$$i_{in} = g_i(v_2 - v_3) + g_B v_2 \qquad (A1)$$

at node 3,

$$g_i(v_2 - v_3) + g_m(v_2 - v_3) = v_3(g_o + g_E + g_L). \qquad (A2)$$

The method is to find v_3 in terms of v_2 from equation (A2), and then substitute for v_3 in equation (A1).

Equation (A2) can be written as

$$v_2(g_i + g_m) = v_3(g_o + g_E + g_L + g_i + g_m)$$

so

$$v_3 = \frac{v_2(g_i + g_m)}{g_o + g_E + g_L + g_i + g_m}. \qquad (A3)$$

Substituting this in equation (A1) gives

$$i_{in} = v_2(g_B + g_i) - \frac{v_2 g_i(g_i + g_m)}{g_o + g_E + g_L + g_i + g_m}.$$

Typically g_B is much less than g_i, so this equation reduces to

$$\frac{i_{in}}{v_2} = \frac{g_i(g_o + g_E + g_L)}{g_o + g_E + g_L + g_i + g_m}. \qquad (A4)$$

In addition if g_o is very much less than g_E or g_L, and $g_i \ll g_m$, the input conductance is

$$\frac{i_{in}}{v_2} \approx \frac{g_i(g_L + g_E)}{g_L + g_E + g_m}$$

or, by dividing through by $(g_L + g_E)g_i$ and inverting, and remembering that $g_m/g_i = \beta$, the input *resistance* is

$$\frac{v_2}{i_{in}} = r_i + \frac{\beta}{g_E + g_L}. \qquad (A5)$$

And if, as is usual, $g_E \ll g_L$,

$$R_{in} = r_i + \beta R_L.$$

(b) *Voltage gain* is given by v_3/v_2. This has already been calculated in equation (A3). That is

$$\frac{v_3}{v_2} = \frac{g_i + g_m}{g_o + g_E + g_L + g_i + g_m}. \tag{A6}$$

This is approximately equal to 1 because $g_o + g_E + g_L$ is normally much less than g_m.

(c) *Output conductance*, is given by $i_{out}/v_{out} = i_{out}/v_3$ with $v_s = 0$. To calculate output conductance you 'look back into' the circuit from the output terminal (with the load resistance removed). The source remains in the circuit but it produces zero signal voltage so $v_s = 0$. The nodal equations are therefore slightly different now; they exclude g_L and include g_S, but v_1 at node 1 is 0 V.

At node 2,

$$(g_S + g_B)v_2 + g_i(v_2 - v_3) = 0. \tag{A7}$$

At node 3,

$$i_{out} = v_3(g_E + g_o) - (g_m + g_i)(v_2 - v_3).$$

or

$$i_{out} = v_3(g_m + g_i + g_E + g_o) - v_2(g_m + g_i). \tag{A8}$$

Rearranging equation (A7) gives

$$v_2 = \frac{v_3 g_i}{g_S + g_B + g_i}.$$

Substituting this value in equation (A8) gives

$$i_{out} = v_3(g_m + g_i + g_E + g_o) - \frac{v_3 g_i(g_m + g_i)}{g_S + g_B + g_i}$$

so

$$\frac{i_{out}}{v_3} = \frac{(g_S + g_B + g_i)(g_m + g_i + g_E + g_o) - g_i(g_m + g_i)}{g_S + g_B + g_i}$$

$$= \frac{(g_S + g_B)(g_m + g_i + g_E + g_o) + g_i(g_E + g_o)}{g_S + g_B + g_i}.$$

Now if, as before, g_i, g_E and g_o are much less than g_m, and g_o is much less than g_E this equation reduces to

$$\frac{i_{out}}{v_{out}} \approx \frac{(g_S + g_B)g_m + g_i g_E}{g_S + g_B + g_i}. \tag{A9}$$

If $g_S + g_B$ is much smaller than g_i (i.e. if the source has a high internal resistance), then, remembering that $g_m/g_i = \beta$,

$$g_{out} = \frac{i_{out}}{v_{out}} \approx \beta(g_S + g_B) + g_E.$$

Thus the source resistance is effectively R_S and R_B in parallel, and the output *resistance* is R_E in parallel with this effective source resistance divided by β.

On the other hand if g_S is much larger than g_i or g_B (i.e. if the source has a low internal resistance), then equation (A9) reduces to

$$\frac{i_{out}}{v_{out}} \approx \frac{g_S g_m + g_i g_E}{g_S}$$

and since $g_i g_E \ll g_S g_m$ the output conductance is

$$\frac{i_{out}}{v_{out}} \approx g_m \text{ or } R_{out} \approx r_e. \tag{A10}$$

So if the signal source has a low internal resistance, the output resistance of the emitter follower is, to a good approximation, simply $r_e = 1/g_m$. For example if the d.c. current is 1 mA the output resistance is about 25 Ω.

INDEX OF KEY TERMS

PART 3   TRANSISTOR SWITCHING CIRCUITS

CONTENTS

AIMS

The aims of this third part of Block 4 are:

1 To describe and explain the switching properties of transistors.

2 To describe the design of various widely used families of transistor switching (or digital) circuits.

3 To explain the properties of a selection of these circuits in terms of the properties of the transistors from which they are made.

OBJECTIVES

GENERAL OBJECTIVES

After studying this text you should be able to:

1 Explain and use correctly the following terms and acronyms:

As regards logic circuits:

TRL, ECL, TTL, STTL, nMOS, standard CMOS, CMOS transmission gates,
fan-in, fan-out, noise margins,
passive pull-up or pull-down, active pull-up or pull-down,
response times: rise time, fall time, delay time, saturation time, propagation delay time, turn-on time, turn-off time,
wired-OR and wired-AND circuits,
load line and load line construction,

As regards transistors and diodes:

base charge (of a bipolar transistor),
gate capacitance, p–n junction capacitance,
saturation base charge (of a bipolar transistor),
saturation region of operation of a bipolar transistor,
Schottky diode, transit time, saturation time constant.

2 Explain the properties of TRL, ECL and CMOS switching circuits in terms of the properties of the devices from which they are made.

3 Describe the structure and the properties of TTL and nMOS.

SPECIFIC OBJECTIVES

When you have completed your study of this text you should be able to:

1 Use the load line construction to establish the operating points of simple circuits.

2 Estimate response times of TRL, ECL and CMOS switching circuits.

1 INTRODUCTION

In addition to being used as an amplifier, as described in Part 2, a transistor can also be used as a switch. That is, it can be regarded as a device which will either pass a current with very little voltage across it, or it will pass almost zero current even when there is a significant voltage across it. A mechanical domestic wall switch illustrates the two important states of an ideal switch: either it is ON and there is virtually zero voltage across it, or it is OFF and there is virtually zero current through it. Although a transistor cannot quite achieve these ideal states—there is a small voltage across it when it is ON, and a small current through it when it is OFF—its ON and OFF states can nevertheless be used to represent the 0 and 1 states in digital logic circuits. In addition, however, transistors have a wide range of advantages in digital circuits as compared with mechanical wall switches. For example: (a) a wall switch is *manually* or *mechanically* controlled whereas a transistor is *electronically* controlled, which means that you can use other circuits to control it; (b) a wall switch is relatively large—each one occupies perhaps a minimum of a cubic centimetre, whereas electronic switches can nowadays be made extremely small— more than 100 000 of them on a square centimetre of a silicon wafer; (c) wall switches can be operated at a maximum speed of perhaps 5 times a second; even the fastest electrically-driven mechanical switches—such as high-speed relays—can operate only a few hundred times a second, whereas transistors can be switched a million or so times faster; (d) wall switches are many times more expensive, and so on. However, unlike mechanical switches, most transistors cannot be expected to handle the voltage and power of the a.c. mains, so each type of switch has its specific use.

This part of Block 4 develops the description of transistors a little further than did Part 1, so that it embraces their switching properties. In particular it explains the properties of a bipolar transistor when it is ON and operating in the saturation region, and it explains the factors which determine how fast transistors can be switched (electronically) from one state to the other.

You will *not* be using your computer to calculate the properties of the various switching circuits because the transistor model used in NODALOU does not include some of the key parameters which affect switching speed. It is necessary to use the more detailed simulation programs, such as SPICE, to carry out such calculations. Your work hitherto with NODALOU is however an excellent preparation for the use of SPICE should you have access to a suitable computer.

As it happens, the mathematics needed to estimate quite accurately the switching speed of logic circuits is very simple—just addition, subtraction, multiplication and division—so you can obtain useful results without a computer to help you; you only need a calculator. Section 2 therefore concentrates mostly on helping you understand what is happening within transistors when they are switched on and off so that you can set up the equations to be solved. The various types of logic gates constructed from transistors are described and explained in Section 3.

SAQ 1 Figure 1(b) shows the d.c. characteristics of a typical n–p–n bipolar transistor. If the transistor is to be used as a switch in a circuit in which the d.c. supply is 6 V, and if the ON current is to be 2 mA, indicate on the diagram suitable operating points for the ON and OFF states of the transistor when it is used as a switch. Deduce what input base currents should be applied to achieve these two operating points.

2 THE SWITCHING PROPERTIES OF TRANSISTORS

2.1 THE LOAD LINE

For a circuit such as that shown in Figure 1(a) there is a useful graphical construction by means of which it is possible to determine the possible operating points of the transistor even when the device is not behaving linearly. This is called the **load line construction**. With its help you can more precisely answer the kind of question posed in SAQ 1.

Figure 1(b) shows again the typical d.c. characteristics of a bipolar transistor, drawn with equal steps of I_B between the different curves. Each point on one of the curves represents a possible operating point of the transistor, expressed in terms of I_C, V_{CE} and I_B.

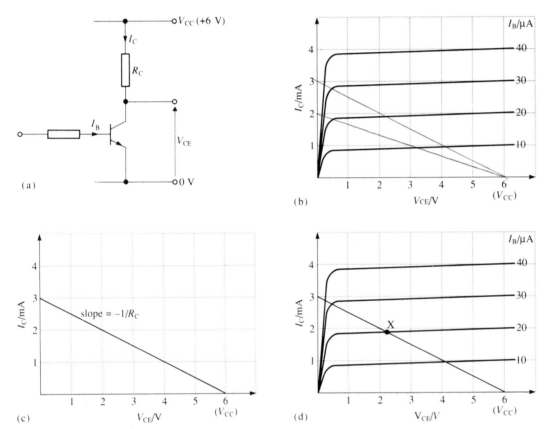

FIGURE 1 (a) The basic common-emitter circuit of a bipolar transistor which can be used as a switch; (b) typical d.c. characteristics of a bipolar transistor; (c) the characteristic of the resistance R_C plotted as a 'load line'; (d) graphs (b) and (c) superimposed to form the 'load line construction' for finding the operating point of the circuit of (a)

☐ Sketch in the curve that would correspond to a base current of 25 µA.

■ It would lie approximately midway between the 30 µA curve and the 20 µA curve. It represents the locus of another possible set of operating points. Indeed every point on the graph represents a possible operating point, but the values of I_B must be inferred by interpolation if the point of interest is not on one of the curves.

Figure 1(c) shows the 'd.c. characteristic' of the resistor R_C in Figure 1(a). The graph is a plot of the voltage at the 'bottom' of this resistor versus the current flowing through it. That is, the voltage plotted is 'V_{CC} minus the voltage drop across the resistor', namely $(V_{CC} - I_C R_C)$. For example, if $I_C = 2$ mA the voltage plotted when $V_{CC} = 6$ V and $R_C = 2$ kΩ is 2 V.

The purpose of plotting the resistor characteristic in this way, instead of just plotting current versus voltage drop, is so that the voltage plotted is the voltage V_{CE} across the transistor. Hence the voltage and current axes of both Figures 1(b) and 1(c) are the same, namely V_{CE} and I_C. This graph of the resistor characteristic is called a **load line**. In this case it is simply a straight line with a slope of $-1/R_C$ passing through the voltage V_{CC} on the voltage axis. The *load line* has a *negative* slope because V_{CE} decreases as I_C increases, though, of course, the *resistor* has a *positive* resistance.

☐   What is the resistance of the load resistor R_C plotted in Figure 1(c)? Sketch on the figure the load line for a load resistance of 4 kΩ in the same circuit.

■   The slope of the line is evidently $-(3 \text{ mA})/(6 \text{ V})$, which means that $R_C = 6 \text{ V}/3 \text{ mA} = 2 \text{ k}\Omega$. The load line for a 4 kΩ resistor is a straight line drawn from $V_{CE} = 6 \text{ V}$ to $I_C = 1.5 \text{ mA}$. That is, its slope is $-1/(4 \text{ k}\Omega)$.

Since Figures 1(b) and 1(c) are both plots of I_C versus V_{CE} they can be superimposed as in Figure 1(d). Then, since an operating point of the circuit of Figure 1(a) must lie on *both* graphs, it must be at a point of *intersection* between them. So, for example, if a base current I_B of 20 μA is supplied to the transistor, you can immediately read from the graph at point X, that $I_C = 1.9 \text{ mA}$ and that $V_{CE} = 2.2 \text{ V}$. Thus the load line construction is a simple way of arriving at the operating point of a nonlinear circuit. (It can equally well be used when the transistor and its load are both nonlinear; see Section 3.6.1.)

SAQ 2	Figure 1(d) is the load line construction for the circuit of Figure 1(a) for a particular transistor when the load resistor R_C has a resistance of 2 kΩ. Find the following:

(i)   the operating point when $I_B = 10 \text{ μA}$;
(ii) the operating point when $I_B = 40 \text{ μA}$;
(iii) the approximate current gain β of the transistor;
(iv) The load resistance needed to achieve the two operating points identified in SAQ 1.

The usefulness of the load line construction where switching circuits are concerned is that it shows how to identify two distinct operating 'areas' which can by used to represent the two binary states 0 and 1. Thus if the 0 state (output low) is represented by $V_{CE} < 0.5 \text{ V}$ you can see in Figure 1(d) that I_B for this transistor with a 2 kΩ load must be greater than 30 μA. If the 1 state (output high) is represented by $V_{CE} > 5 \text{ V}$, then I_B must be less than 5 μA.

The low voltage ON state, or 0 state, of the transistor is referred to as the **saturation region of operation*** of the transistor. It refers to the range of operating points at which the collector-emitter voltage is less than the collector-base voltage (i.e. $V_{CE} <$ about 0.65 V). However, before I describe this region of operation in some detail, I would like you to consider a similar circuit to that of Figure 1(a), but which uses an enhancement-mode n-channel MOSFET, as shown in Figure 2(a).

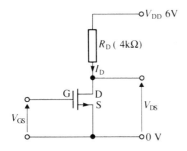

(a)

FIGURE 2   (a) The basic common-source circuit of a MOSFET

*The saturation region of operation of a bipolar transistor is not the same as that of a MOSFET. It is unfortunate that the same term is used to describe quite different domains of operation in the two types of transistor, but one just has to accept it.

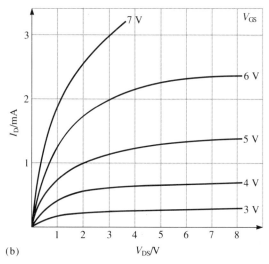

(b)

FIGURE 2 (b) The d.c. characteristics of an enhancement-mode, n-channel MOSFET

> **SAQ 3**
>
> Figure 2(b) shows a typical family of d.c. characteristics of an enhancement-mode n-channel MOSFET. Draw the load line that applies to the circuit of Figure 2(a) on Figure 2(b) and determine the input voltages V_{GS} needed to establish two distinct output voltages V_{DS} that could be used to represent the digital values 0 and 1.
>
> Note that the two output voltages you arrive at should be suitable to be used as input voltages for a subsequent identical switching circuit (i.e. they must also be suitable values of V_{GS}).

You should have found in SAQ 3 that the output voltage when a MOSFET is in the ON state, is a good deal larger in this circuit than it is using a bipolar transistor, which implies more power dissipation and a less well-defined ON state. Partly for this reason, but mainly because resistors are more expensive to make in integrated form than transistors, the circuit of Figure 2(a) is not used in digital circuits (see Section 3.6 for more practical MOSFET circuits).

2.2 BIPOLAR TRANSISTORS IN SATURATION

The saturation region of operation of a bipolar transistor is usually used as the ON state of a switching transistor. It refers to the state in which current passes through the transistor and the voltage drop across it is very small. It is the region of Figure 1(b) in which the collector-emitter voltage is less than the base-emitter voltage (i.e. when $V_{CE} < V_{BE}$). It is the region in which the output characteristics are very curved, and it is the region in which I_B exceeds I_C/β.

Consider what happens in the circuit of Figure 3 when I_B is increased.

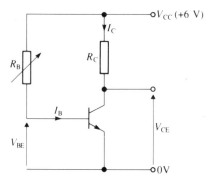

FIGURE 3 A common-emitter circuit with variable base current drive. The edge of the *saturation* region is when $V_{CE} = V_{BE}$ and when the collector current is $I_{C(max)} = (V_{CC} - V_{BE})/R_C$. The transistor is in the *saturation region of operation* when $I_B > I_C/\beta$, i.e. when $V_{CE} < V_{BE}$. In these circumstances V_{CE} is called $V_{CE(sat)}$. The more I_B exceeds $I_{C(sat)}/\beta$ the smaller $V_{CE(sat)}$ becomes. Typically, in saturation, $V_{BE} \approx 0.7$ V and $V_{CE(sat)}$ is about 0.2 V

Starting with the resistance of the variable resistor R_B very large—and I_B very small—the output voltage is approximately equal to V_{CC} (see Figure 1(d)). As R_B is reduced, I_B and therefore I_C as well, are increased, causing V_{CE} to decrease (i.e. the operating point moves to the left along the load line in Figure 1(d).) At a certain value of R_B the collector current will be such that the collector voltage will be the same as the base voltage; that is $V_{CE} = V_{BE} \approx 0.65$ V, or $V_{CB} = 0$. *This is the edge of the saturation region.* (Remember, as explained in Part 1, $I_C = \beta I_B$ even when there is zero voltage across the collector-base junction.)

A further increase in I_B drives the transistor into saturation. That is I_C still increases a little and so forces V_{CE} to be *less* than V_{BE}. I_C simply gets closer and closer to V_{CC}/R_C as I_B increases. From the edge of saturation, when $V_{CE} \approx 0.65$ V to the fully saturated state when $V_{CE} = $ (say) 0.1 V the collector current only increases by a few per cent even if I_B doubles or trebles. So, in saturation, $I_B > I_C/\beta$.

Notice that in saturation the collector-base p–n junction acquires a *forward voltage* bias even though there is still a *reverse current* flowing through it. Since V_{CE} falls to perhaps 0.1 V whilst V_{BE} stays much the same, the p-type base region becomes more positive than the n-type collector region by about 0.55 V, and the junction is forward biased. Yet the current still flows from the n-region towards the p-region—the direction of a reverse current! (Evidently forward biasing a p–n junction does not necessarily produce a forward current, as explained in Part 1.)

The voltage between the collector and the emitter when the transistor is saturated is called the **saturation voltage** $V_{CE(sat)}$. Typically $V_{CE(sat)}$ is between 0.1 V and 0.2 V for a low-power transistor. You can see from Figure 1(d) that the larger you make the base current the smaller will be $V_{CE(sat)}$ for a given load resistor. In many switching circuits (but not in ECL, see Section 3.5.2) these two values of V_{CE} (namely $V_{CE} = V_{CE(sat)}$ and $V_{CE} \approx V_{CC}$) are used to represent the logic values 0 and 1 in digital circuits.

| SAQ 4 | In the circuit of Figure 3, in which the transistor has a current gain of about 100, R_B is adjusted so that $I_B = 40$ μA. If $R_C = 3$ kΩ, calculate: |

(i) $V_{CE(sat)}$ (draw your load line on Figure 1(b)),
(ii) I_C,
(iii) the resistance of R_B,
(iv) the ratio I_C/I_B when the transistor is driven into saturation.

2.3 BASE CHARGE IN BIPOLAR TRANSISTORS

You might well be wondering what happens to this extra base current flowing into the transistor when the transistor is saturated. Hitherto I have only explained how it is that base current gives rise to a collector current which is β times as big when the collector is reverse biased. In saturation, however, I_B increases well beyond I_C/β because I_C cannot exceed V_{CC}/R_C. So what happens to the extra base current?

The answer is that it causes extra charge to accumulate in the base region. The normal electron density profile, as explained in Part 1, is indicated in Figure 4(a). The charge in the base region, called **base charge** Q_B as shown,

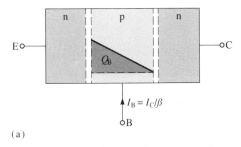

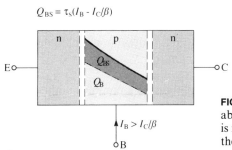

FIGURE 4 Charge density distributions in the base region of a transistor, over and above the equilibrium density: (a) in normal operation, when the collector junction is reverse biased and the emitter junction is forward biased, a charge of Q_B exists in the base region; (b) when the transistor is saturated an extra 'saturation charge' Q_{BS} accumulates in the base region

actually consists of equal amounts of holes and electrons in excess of the doped-in carriers which are always there. In an n-p-n transistor the base current supplies the holes whilst the electrons enter the base region via the emitter junction. The magnitude of Q_B is simply the current I_C flowing through the base region from emitter to collector, multiplied by the time it takes to do so, namely the **transit time** τ_t. That is,

$$Q_B = I_C \tau_t. \tag{1}$$

When the transistor is driven into saturation the collector current cannot increase further so the electrons drawn into the base by the base current of holes are forced to stay in the base region, causing an additional charge to build up there. This additional charge is called the **saturation base charge** Q_{BS}, and is shown diagrammatically in the Figure 4(b) by the area labelled Q_{BS}. This saturation base charge also consists of equal quantities of holes and electrons.

However, as explained in Part 1, holes and electrons tend to recombine so Q_{BS} does not go on increasing indefinitely. A point is soon reached at which the holes and electrons comprising Q_{BS} recombine at a rate which equals the extra base current being supplied. This then is the steady-state value of the saturation charge Q_{BS}. The smaller the recombination rate of this saturation charge (i.e. the longer its lifetime) the greater is the steady-state value of Q_{BS} for a given excess base current. Thus if τ_s is the lifetime of the carriers comprising Q_{BS}, then, in the steady state, Q_{BS} is equal to τ_s times the excess base current that is driving the transistor into saturation. Hence

$$Q_{BS} = \tau_s(I_B - I_C/\beta). \tag{2}$$

It turns out, as explained later, that this saturation base charge can be quite troublesome when fast switching speeds are required, so *short* lifetime material is desirable if Q_{BS} is to be kept small.

☐   Why is long lifetime material desirable for *normal* operation of a transistor?

■   Because it means that very few of the minority carriers that are flowing from the emitter to the collector recombine in the base region, so I_B is small and β is large. There are thus conflicting requirements on the lifetime of carriers in the base region.

In summary then:

(i) the amount of charge in the base region of a transistor which is conducting normally (i.e. when its collector is reverse biased) is Q_B, where

$$Q_B = I_C \tau_t. \tag{1}$$

τ_t, the transit time, is the time minority carriers (holes in p–n–p transistors, electrons in n–p–n ones) take to cross the base region from emitter to collector;

(ii) the additional base charge, Q_{BS}, which accumulates in the base region of a saturated transistor depends on

  (a) the base current in excess of I_C/β, namely $I_B - I_C/\beta$,

  (b) the lifetime τ_s of the saturation charge in the base region.

So

$$Q_{BS} = \tau_s(I_B - I_C/\beta). \tag{2}$$

τ_s is alternatively called the **saturation time constant** and is sometimes quoted in transistor data sheets. Its magnitude might be $50 \times \tau_t$ or more, so, even for a small base current in excess of I_C/β, Q_{BS} may be many times as big as Q_B.

SAQ 5 When a transistor is driven into saturation it is evident that: (i) the extra base current causes the emitter current to increase, and (ii) the collector junction becomes forward biased. How are these two facts represented in the diagram of Figure 4(b)? From Figure 4(b) what do you deduce happens to V_{BE} when the transistor is driven into saturation?

2.4 THE SWITCHING TIMES OF TRANSISTORS

When a transistor is switched from the OFF state to the ON state and back again it doesn't change state immediately; it takes a finite time to respond. For example, when the square-wave voltage shown in Figure 5(b) is applied to the input of the circuit of Figure 5(a), a typical output waveform is that shown in Figure 5(c). In this case R_B has been chosen so that $V_S/R_B > (V_{cc}/R_c)/\beta$, which means that the base current when the input is high is sufficient to drive the transistor into saturation. You can see that the transistor's response to the abrupt changes at the input is much slower; although, to keep things in perspective, it is important to appreciate that in good circuit designs the times t_{on} and t_{off} are likely to be of the order of only a few tens of nanoseconds at most.

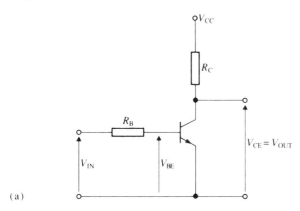

(a)

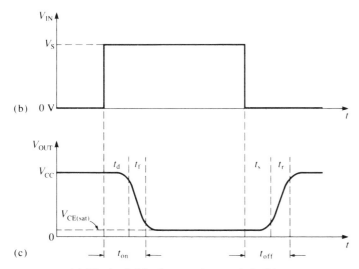

(b)

(c)

FIGURE 5 (a) The basic bipolar transistor switch; (b) a square-wave voltage input waveform; (c) the resulting output waveform, showing turn-on time t_{on}, turn-off time t_{off}, delay time t_d, fall time t_f, saturation time t_s and rise time t_r.

The **turn-on time** t_{on} is the time from the instant that the input voltage goes positive to the moment when the output voltage has fallen almost to $V_{CE(sat)}$ (the level usually specified is 10% of $V_{CC}-V_{CE(sat)}$.) Turn-on time is then divided into a **delay time** t_d and a **fall time** t_f as indicated in the figure.

Similarly the **turn-off time** t_{off} is the time from the moment that the input voltage returns to zero to the moment when the output has risen to (90% of) its final value. Turn-off time is usually subdivided into the **saturation time** t_s and the **rise time** t_r, as indicated in the diagram. The 10% and 90% figures are introduced to give a precise instant at which to measure these times; the moment at which they reach their final values is never very clear. (For simplicity I shall ignore the 10% and 90% figures in the following explanations and assume that the transistor is switched off and on completely). What causes these finite **response times**?

In general, the speed at which a device can be switched on depends upon the amount of current available to provide the charge it needs as it changes from the OFF state to the ON state. And the rate at which it can be switched off depends on the current available to remove the charge again. Thus if Q_{ON} is the total charge required to turn on a transistor, and if $I_{B(ON)}$ is the (average) current available to do so, then, for a fairly fast turn-on, the turn-on time is given by

$$t_{on} \approx \frac{Q_{ON}}{I_{B(ON)}} \tag{3}$$

Similarly, if Q_{OFF} is the charge needed to turn the transistor off, the turn-off time is given by

$$t_{off} \approx \frac{Q_{OFF}}{I_{B(OFF)}} \tag{4}$$

Of course $I_{B(ON)}$ and $I_{B(OFF)}$ must flow in opposite directions since they carry charge into or away from the transistor. For simple calculations they have to be thought of as average values because they may well change somewhat during the switching process. Similarly, Q_{ON} may not be equal to Q_{OFF}, as I shall explain, so each quantity in equations (3) and (4) has to be known before the response times of a device in a particular circuit can be calculated.

With bipolar transistors two kinds of charge are involved. First, there is the base charge Q_B, to which Q_{BS} is sometimes added; second, there is the charge needed by the capacitances associated with the emitter and collector p–n junctions as the voltage across them changes ($Q = C \times \Delta V$).

Figure 6 shows the transistor equivalent circuits of transistors that were introduced in Part 2, extended to include the capacitances inherent in the devices. These extended circuits are discussed in more detail in Block 8, Part 1 Here it is only necessary to note that there are capacitances associated with the transistors' input terminals which have to be charged and discharged when the device is switched. The MOSFET has a gate which forms a capacitance with both the source and the drain; the bipolar transistor has base-emitter and base-collector junction capacitances. The voltages across these capacitances change as the transistors are switched, so charge has to be supplied to them at turn-on and removed again at turn-off. These capacitances are actually somewhat nonlinear, but here I shall assume that they have fixed values.

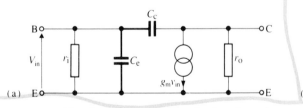 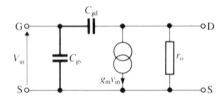

FIGURE 6 The simplified equivalent circuits of transistors extended to include the capacitances. (a) for the bipolar transistor: C_c is the capacitance of the collector-base junction and C_e represents the capacitance of the emitter-base junction; (b) for the MOSFET: C_{gd} is the effective gate-drain capacitance and C_{gs} is the effective gate-source capacitance

The response times shown in Figure 5(b) can be understood in detail simply by considering the times taken to supply and remove the charge the transistor requires.

2.5 THE SWITCHING TIMES OF BIPOLAR TRANSISTORS

Figure 7 is a repeat of Figure 5 except that the input waveform is drawn as a current waveform which changes abruptly from $I_{B(ON)}$ to $I_{B(OFF)}$ and back again as the input square-wave voltage changes from high to low and back again.

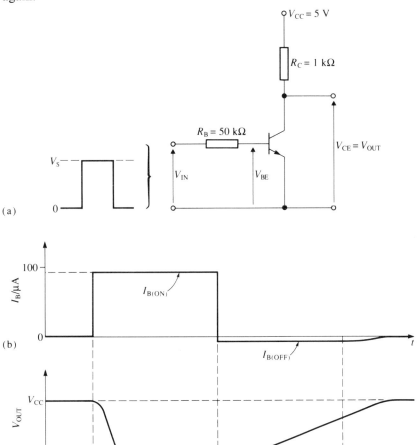

FIGURE 7 (a) A specific example of the circuit of Figure 5(a) in which: current gain $\beta = 200$, transit time $\tau_t = 0.4$ ns, saturation time constant $\tau_s = 20$ ns, emitter capacitance $C_e \approx 2$ pF, collector capacitance $C_c \approx 1.5$ pF and $V_{CE(sat)} \approx 0.2$ V; (b) the input current waveform; (c) the output waveform

To begin with assume that V_{IN} has been at zero volts for some time so that the transistor is fully cut-off, and that $V_{BE} = 0$ V, $I_B = 0$, $I_C = 0$ and $V_{OUT} = 5$ V. This is the condition at $t = 0$ in Figures 7(b) and (c).

When V_{IN} goes high to V_S, base current $I_{B(ON)}$ immediately starts to flow into the transistor. This causes the base voltage to start to rise. It increases to about 0.65 V before any collector current flows because this is the forward bias needed to produce a significant current through the emitter p–n junction. The base voltage doesn't rise instantaneously because of the charge required by the capacitances C_e and C_c shown in Figure 6. The voltage across both these junctions changes by 0.65 V, so the total charge required is $(C_e + C_c) \times 0.65$ V. *The time taken to supply this charge is the delay time t_d in the output waveform (see Figure 5).*

During this time the input current decreases a little as V_{BE} rises, but a constant value of $I_{B(ON)} = (V_S - V_{BE})/R_B$, as shown in Figure 7(b), is usually accurate enough for estimating delay times.

Following the delay time, the base current stays at $(V_S - V_{BE})/R_B$ and the collector current starts to flow. The output voltage therefore begins to fall.

Again it doesn't fall immediately because $I_{B(ON)}$ has to supply more charge to the collector capacitance as well as supply the base charge Q_B. *The time it takes for $I_{B(ON)}$ to supply these charges is the fall time t_f.* If $V_{CE(sat)} \approx 0.2$ V is the final output voltage when the transistor is ON, the change of voltage across C_c is $V_{CC} - V_{CE(sat)}$, which is about 4.8 V in this case. If $I_{C(ON)}$ is the final collector current, the base charge Q_B required is $\tau_t I_{C(ON)}$, (see equation (1)).

Note that whilst V_{IN} remains at V_S, base current continues to flow, and since R_B has been chosen so that $I_{B(ON)} > I_C/\beta$, saturation charge Q_{BS} accumulates in the base region, as previously explained.

At turn-off, when V_{IN} abruptly falls to zero volts again, the base current changes direction and becomes $I_{B(OFF)} = V_{BE}/R_B$ because V_{BE} does not immediately fall to zero. The first thing that this reversed base current has to do is remove the saturation charge Q_{BS}. *The time this takes is the saturation time $t_s = Q_{BS}/I_{B(OFF)}$.* Throughout this time I_C and I_E continue to flow because the normal base charge Q_B is still in the base region, which is why V_{BE} does not yet fall to zero again.

Finally, after Q_{BS} is exhausted, the collector capacitance is discharged again and the base charge Q_B is removed—causing I_C to decrease in proportion. *The time this takes is the rise time t_r.* Although the charge involved here is the same as for the fall time, the current $I_{B(OFF)}$ is not the same as $I_{B(ON)}$ so $t_r \neq t_f$. This completes the turn-off process.

Only after the collector current has stopped flowing do V_{BE} and I_B fall again to zero.

Note that, with bipolar transistors the saturation charge Q_{BS} has to be *removed* before the collector current starts to fall, but it does not have to be *supplied* during turn-on in order to bring the output voltage to near zero volts. Q_{BS} builds up fairly slowly *after* the transistor has been turned on. But it has to be removed rapidly if the transistor is to be turned off quickly.

Rapid switching is one of the main aims in the design of digital circuits—especially for computers—so the question of how switching circuits can be made faster is always present. One might suppose that, in order to achieve shorter switching times in the circuit of Figure 7(a), the thing to do would be to decrease R_B and so increase both $I_{B(ON)}$ and $I_{B(OFF)}$. In practice this usually has the effect of decreasing t_d, t_r, and t_f as intended, but of *increasing* t_s. This is because the larger $I_{B(ON)}$ increases Q_{BS} by a greater factor than the rate at which the larger $I_{B(OFF)}$ can remove it again. This is illustrated in the following calculations. So Q_{BS} has to be given special attention in the design of high-speed circuits.

2.5.1 ESTIMATING SWITCHING TIMES OF A BIPOLAR TRANSISTOR

All the above explanations can quite simply be translated into numerical values as follows.

In Figure 7(a) suppose that $V_S = V_{CC} = 5$ V and that $R_B = 50$ kΩ and $R_C = 1$ kΩ. The problem is to calculate the response times given the performance parameters of the transistor.

Let us suppose that the transistor parameters are as given in the caption to Figure 7. With these data the response times can now be calculated:

(a) First calculate $I_{B(ON)}$ and $I_{B(OFF)}$:

Since $R_B = 50$ kΩ it follows that $I_{B(ON)} = \dfrac{5\ \text{V} - 0.65\ \text{V}}{50\,000\ \Omega} = 87\ \mu\text{A}$,

and that $I_{B(OFF)} = \dfrac{0.65\ \text{V}}{50\,000\ \Omega} = 13\ \mu\text{A}$

(b) Next calculate Q_B and Q_{BS}:

With $R_C = 1$ kΩ, and $V_{CE(sat)} = 0.2$ V, it follows that

$$I_{C(ON)} = \frac{4.8 \text{ V}}{1 \text{ k}\Omega} = 4.8 \text{ mA}.$$

Therefore (by equation (1)) $Q_B = I_{C(ON)} \tau_t = 4.8$ mA $\times$ 0.4 ns $= 1.92$ pC.

Similarly (by equation (2)) $Q_{BS} = \tau_s(I_{B(ON)} - I_{C(ON)}/\beta)$

$$= 20 \text{ ns} \times (87 \text{ μA} - 4.8 \text{ mA}/200)$$

$$= 1.26 \text{ pC}.$$

(c) It is now possible to calculate the response times

$$Delay \ time \ t_d \approx \frac{(C_c + C_e) \times 0.65 \text{ V}}{I_{B(ON)}} = \frac{3.5 \text{ pF} \times 0.65 \text{ V}}{87 \text{ μA}} = 26 \text{ ns}. \qquad (5)$$

$$Fall \ time \ t_f \approx \frac{Q_B + C_c \times \Delta V_{CE}}{I_{B(ON)}} = \frac{1.92 \text{ pC} + 1.5 \text{ pF} \times 4.8 \text{ V}}{87 \text{ μA}} = 105 \text{ ns}. \quad (6)$$

$$Saturation \ time \ t_s \approx \frac{Q_{BS}}{I_{B(OFF)}} = \frac{1.26 \text{ pC}}{13 \text{ μA}} = 97 \text{ ns}. \qquad (7)$$

$$Rise \ time \ t_r \approx \frac{Q_B + C_c \times \Delta V_{CE}}{I_{B(OFF)}} = \frac{1.92 \text{ pC} + 1.5 \text{ pF} \times 4.8 \text{ V}}{13 \text{ μA}} = 702 \text{ ns}. \ (8)$$

Hence $t_{on} = t_d + t_f = 131$ ns $\qquad\qquad\qquad\qquad\qquad (9)$

and $t_{off} = t_s + t_r = 799$ ns. $\qquad\qquad\qquad\qquad\qquad (10)$

Notice how much longer the turn-off time is compared with the turn-on time, mainly as a result of the small value of $I_{B(OFF)}$. With a rise time as long as this, recombination of the carriers in the base region may significantly help with the removal of base charge and so produce a faster response than that calculated.

> **SAQ 6** Calculate the six response times listed in the above example that result from reducing the resistance of R_B in Figure 7(a) to 30 kΩ.

As you will see later, shorter response times than those calculated above can be achieved with the same transistor using better circuit designs such as TTL and ECL which will be described in Section 3.5. This example, however, serves to bring out the concepts that have to be understood when designing good digital gates or in estimating their performance.

In all the above calculations of rise and fall times it has been assumed that there is negligible capacitive load on the output of the transistor. In practice of course there always is some capacitance connected to the output, even if it is only the stray capacitances due to the wiring. Whether or not these capacitances affect these response times significantly depends on their magnitude, on the current available to charge and discharge them and on the inherent response times of the transistor circuit.

☐ Why don't the stray capacitances connected to the output of a transistor affect the delay time and the saturation time of the transistor response?

■ Because a capacitive load only affects *changes* in output voltage. During t_d and t_s the output voltage does not change so the capacitive load draws no current from the output.

2.5.2 THE EFFECT OF SPEED-UP CAPACITORS

One method that can be used to obtain faster response times in the circuit of Figure 7(a) is to insert a speed-up capacitor C_B in parallel with R_B as shown in Figure 8 (overleaf). Insertion of a capacitor in this way is not always possible, especially in integrated circuits where capacitors take up so much surface area, and they load the driving circuit severely. However, if a good square-wave voltage source is available the insertion of a speed-up capacitor in parallel with R_B ensures much faster response times. Why this is so can be understood as follows.

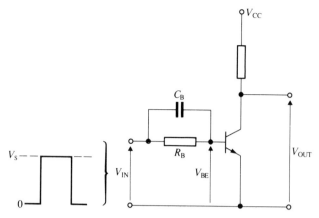

FIGURE 8 The circuit of Figure 7(a) with a 'speed-up' capacitor C_B added

To turn a bipolar transistor on so that a steady current flows through it, it is necessary to provide both the base charge Q_B (including the charge required to charge C_e and C_c) and the base current I_B; the charge is needed to *establish* the collector current and the base current is needed to *sustain* it. In the circuit of Figure 7(a) the base current through R_B has to do both jobs: it has to supply the base charge, but as soon as it has begun to do so, some of the base current is taken up in sustaining it as it begins to recombine. Hence Q_B and the collector current build up exponentially. The behaviour is rather like the process of charging a capacitor via a resistor described in Block 2, Part 1.

In the circuit of Figure 8, however, the input capacitor C_B deals with the base *charge* so that the resistor R_B only has to supply the d.c. *current* to sustain the charge, namely I_C/β. Thus if the input voltage changes abruptly from 0 V to V_S the transistor will be turned on almost instantaneously if C_B supplies the required turn-on charge Q_{ON}; that is if

$$C_B = \frac{Q_{ON}}{V_S - V_{BE}}. \tag{11}$$

The rise time will be almost as fast as the input step of voltage driving the circuit. The inherent delay introduced by the transistor is only its transit time τ_t which, in the above example, is 0.4 ns!

The base current I_C/β through R_B is needed to sustain this charge, as explained above, so, if the transistor is not driven into saturation,

$$R_B = \frac{V_S - V_{BE}}{I_C/\beta}. \tag{12}$$

With these values of R_B and C_B the transistor will turn on and off abruptly in response to an input square wave of amplitude V_S. If R_B is smaller than the value given in equation (12), the transistor will be driven into saturation and a larger value of C_B will be needed to turn it off abruptly, because Q_{OFF} becomes greater than Q_{ON} due to Q_{BS} building up.

The fact that transistors can in fact be switched so rapidly provides a target to aim for when designing circuits that do not include speed-up capacitors. ECL is one design that very nearly achieves it.

The circuit of Figure 8, but with variable values of R_B and C_B provides a convenient means of measuring Q_B and Q_{BS}. It is only necessary to adjust the values of C_B to give good, square output waveforms, for different values of I_C and I_B, as established by R_C and I_B, to measure Q_B and Q_{BS} and hence calculate τ_t and τ_s, etc. Manufacturers' data sheets usually show the measuring circuits they have used to obtain the turn-on and turn-off times at specified operating points. To give response times at other operating points you have first to calculate τ_t and τ_s and then go through the estimating procedure explained in Section 2.5.1.

SAQ 7 A transistor data sheet gives the saturation time t_s as 50 ns in the circuit of Figure 7(a). What is the saturation time constant τ_s? (You will need to make use of equations (2) and (7) in this calculation.)

2.5.3 THE USE OF SCHOTTKY DIODES

The best way to reduce the saturation base charge in a bipolar transistor switching circuit is to connect a Schottky diode across the collector-base junction as shown in Figure 9(a).

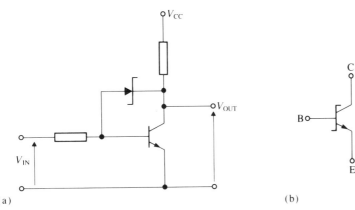

FIGURE 9 (a) A bipolar transistor switch with a Schottky diode connected across the transistor's collector junction to prevent excessive saturation; (b) the usual symbol for this combination of transistor and Schottky diode

A Schottky diode has the same d.c. characteristic as a p–n junction except that the saturation current I_S is much larger, perhaps a million times larger, so that a significant forward current flows through it when the forward bias voltage is much less than for a p–n junction.

Schottky diodes are formed when certain metals, such as aluminium, make very intimate contact with n-type silicon. They can therefore conveniently be made during the production of integrated circuits, as described later. Their mode of operation is quite different from that of p–n junctions but their characteristic equation has just the same form, namely

$$I_D = I_S \left[\exp\left(K V_D\right) - 1 \right]. \tag{13}$$

SAQ 8 Calculate the forward voltage across a Schottky diode whose saturation current I_S is 10^{-9} A when the current flowing through it is 1 mA. ($1/K = 25$ mV.)

So when a Schottky diode is connected across the collector-base junction of a saturated transistor as in Figure 9(a), the *forward* bias of the collector junction is limited to about 0.35 V. (Remember that the collector junction of a saturated transistor becomes forward biased due to the accumulation of saturation base charge, as indicated in Figure 4(b).) And since the charge density in the base region next to the collector junction is an exponential function of the forward bias across the junction, Q_{BS} is greatly reduced. For example, if the forward bias of the collector junction is reduced by only 0.1 V, then $K V_D$ is reduced by the factor 100 mV/25 mV = 4; so Q_{BS} is reduced by the factor $e^4 \approx 55$.

With the Schottky diode connected, the base current which previously flowed into the base region and gave rise to Q_{BS} is now diverted by the forward-biased diode directly into the collector—once V_{BE} is more positive than V_{CE} by about 0.35 V.

Figure 9(b) shows the symbol usually used to indicate transistors with Schottky diodes across their collector junctions. It appears again in the circuit diagram for TTL in Figure 17(b).

2.6 THE SWITCHING TIMES OF MOSFETs

The first point to emphasize with MOSFETs is that they exhibit no effects comparable to the accumulation of saturation charge in bipolars. The d.c. gate current is always zero so charge does not accumulate in the channel if the transistor is over-driven. Gate current does however flow during the process of switching as the gate capacitances are charged and discharged.

So equations (3) and (4), namely $t_{on} \approx \dfrac{Q_{ON}}{I_{B(ON)}}$ and $t_{off} \approx \dfrac{Q_{OFF}}{I_{B(OFF)}}$ still apply, but with I_B being replaced by I_G. The calculation of the open-circuit response times of a MOSFET is therefore mainly one of estimating the gate currents available to switch it on and off.

First consider the input capacitances. Figure 10(a) is a diagram of the construction of a MOSFET, showing in particular the gate electrode of width W, and indicating the length L of the channel. Figure 10(b) indicates the location of the components of the input capacitances. The gate-source capacitance C_{gs} is mainly due to the small overlap between the gate electrode and the source region. Similarly the gate-drain capacitance C_{gd} is due to the overlap with the drain region. The magnitude of the gate-channel capacitance C_{gc} is proportional to the area $W \times L$ of the channel. C_{gc} is distributed along the channel, so it is not obvious how much of it is associated with the source and how much of it is associated with the drain. Actually the appropriate proportions vary with the application, but in order to estimate response times it is necessary to use some simple model. In practice, assuming that about two thirds of C_{gc} is in parallel with C_{gd}, and one third is in parallel with C_{gs} gives about the right answers in typical switching circuits. Again, as a first approximation, the capacitances can be assumed to be linear, so that the input charge associated with each capacitor is simply the product of its capacitance at zero applied voltage and the change of voltage across it during turn-on or turn-off.

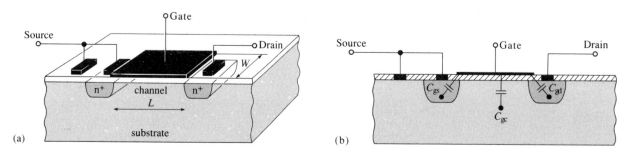

FIGURE 10 Cross-sectional diagrams of a MOSFET: (a) showing the metal gate and the metallic connections through the oxide to the source, substrate and drain; (b) indicating the principal resulting capacitances

> **SAQ 9** Figure 11(a) shows an enhancement-mode MOSFET driven by a square-wave current-source producing turn-on and turn-off currents of 400 μA. Calculate the fall time, delay time and rise time given that, for the transistor involved: $C_{gs} = C_{gd} = 0.2$ pF, $C_{gc} = 1.2$ pF (see Figure 10(b)) and the threshold voltage is 1.5 V. Assume that these capacitances are constant and independent of voltage and that the output voltage falls from 5 V to 1 V during turn-on, and back again during turn-off.

In practice of course MOSFETs are not driven by constant-current sources. As you will see in Section 3.6, a common driving circuit is that formed by T1 and T2 of Figure 11(b). This circuit is the CMOS amplifier which was described in Part 2, except that here it is made from enhancement-mode transistors so that one or other transistor is always cut off. It is explained in more detail later. All I want you to notice at the moment is that the current driving T3 towards the ON state is the drain current of T2 (when T1 is cut-off), whilst the current driving T3 towards the OFF state is the drain current of T1 (when T2 is cut-off). Assuming there is no load applied to the output of T3, its response times depend

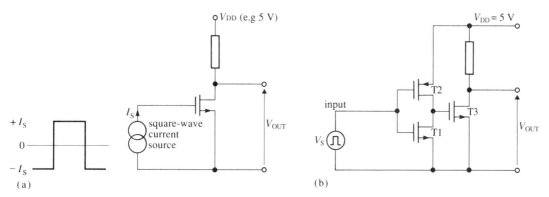

FIGURE 11 (a) The MOSFET version of the basic transistor switch. Note that the input is connected directly to the MOSFET input. (b) A CMOS inverter being used to drive the transitor switch T3.

on these driving currents. Their magnitudes can be derived from a careful analysis of the transistor's nonlinear d.c. characteristics, but this is beyond the scope of this course.

In practice because MOSFET capacitances are so small, and because the transconductance of MOSFETs is much less than that of bipolars (giving smaller output currents), external circuit capacitances to ground usually have a greater effect in MOSFET circuits than in bipolar ones. Their effect is simply to cause the output voltage of a circuit to change more slowly than the changes in output current. The actual differences in response times caused by these external capacitances are rather difficult to estimate because the current changes do not follow a well defined waveform. Circuit simulation programs such as SPICE have to be used to obtain accurate estimates of response times in these circumstances. The effects of loading capacitances will not therefore be further discussed in this course.

☐ *Revision.* What is the capacitance between the metallic strip of a conducting line deposited on silicon dioxide and the silicon substrate beneath it, connecting the output of one transistor to the input of another? The strip is 2 mm long and 6 μm wide and is separated from the silicon substrate by a layer of oxide that is 0.5 μm thick. (The relative permittivity μ_r of silicon dioxide is 4.)

■ The capacitance is given by $C = \dfrac{\mu_o \mu_r \times \text{area}}{\text{thickness of the dielectric}}$.

So $C = \dfrac{8.85 \text{ pF m}^{-1} \times 4 \times 2 \text{ mm} \times 6 \text{ μm}}{0.5 \text{ μm}} = 0.85 \text{ pF}$.

2.7 PROPAGATION DELAY TIMES

The response times discussed hitherto express the behaviour of a transistor switch when it is being driven by a good square wave. In practical digital circuits however, such abrupt transitions from OFF to ON and back again are not normally available. Indeed in well designed digital circuits the input and output voltage transients are similar in shape, both having finite rise and fall times. Consequently it is often not easy to apply the response times discussed hitherto to typical logic circuits and gates. So a simpler and more direct method is often used.

Figure 12 (overleaf) shows the input and output waveforms of an inverting gate, such as a NAND gate. The delay times between the rise and fall times of the two waveforms, indicated by t_{PHL} and t_{PLH}, are called the **propagation times** or **propagation delay times**. They refer to the times between the 50% points of the input and output waveforms. The delay is not always the same at both transients, so two propagation delay times are defined: t_{PHL} is the propagation delay time of the output as it goes from a 'High voltage to a Low voltage', and t_{PLH} is the time for the opposite transition.

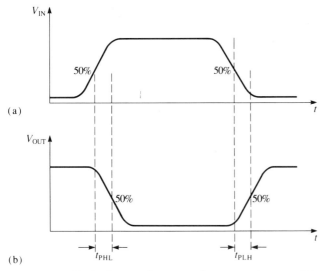

FIGURE 12 (a) The input voltage waveform to an inverter; (b) the corresponding output waveform. The delay between input and output is called the propagation delay time

To a first approximation, the high-low propagation time, t_{PHL}, is equal to $t_{on} = t_d + t_f$, and $t_{PLH} \approx t_{off} = t_s + t_r$, and just t_r for MOSFETs. You will usually find that the speed of *logic gates* is specified in terms of propagation delays, whereas data sheets for *switching transistors* on their own, (i.e. not in any particular circuit) usually give information on the delay times, fall times, saturation times and rise times.

3 FAMILIES OF DIGITAL CIRCUITS (logic families)

3.1 INTRODUCTION

In Block 3, Part 2, a brief description was given of a few TTL and compatible ECL and MOS devices to give you a flavour of the logic families available. Some of the terms considered in this section (e.g. fan out) were introduced in Block 3 from the device *user's* point of view. Now I want to explain them in more detail from the device *designer's* perspective.

Digital circuits are nowadays almost always produced in integrated-circuit form, with at least several gates or flip-flops on one silicon chip. Indeed complete circuits, such as the microprocessor in your microcomputer or large memories involving many thousands of flip-flops, may be present on a single chip. These digital circuits are constructed from one of several possible basic designs, called 'families of digital circuits'. The remaining sections are concerned with describing and explaining some of these circuit families.

The families most widely used in integrated circuits are:

Using bipolar transistors:

TTL (transistor–transistor logic) which is available in several versions;

ECL (emitter-coupled logic) which is based on long-tailed pairs.

Using MOSFETs:

nMOS (n-channel MOS) which uses only n-channel MOSFETs;

CMOS (complementary MOS) which uses pairs of n-channel and p-channel enhancement-mode MOSFETs.

Other families have been designed such as RTL (resistor–transistor logic), DTL (diode–transistor logic), DCTL (direct-coupled transistor logic), CML (current-mode logic, which is similar to ECL), I^2L (integrated-injection logic which uses a crafty combination of an n–p–n and a p–n–p transistor in which the collector region of the p–n–p transistor is also the base region of the n–p–n one), and BiCMOS (which comprises gates with CMOS input and bipolar output). In this course only ECL and CMOS are explained in detail. Briefer descriptions are given of TTL and nMOS since, although they are widely used, their analysis is rather too complicated for this introductory course (see references at the end of this text).

To begin with however I shall consider an early design of transistor logic called TRL (transistor–resistor logic) since it illustrates some general properties of gates and demonstrates some of the problems that the various families of logic circuits are designed to overcome.

(Note that TRL is not to be confused with RTL (resistor–transistor logic) which is still available on the market. The circuit of a 3-input NOR gate in RTL is shown in Figure 13(c); it is not one of the circuits discussed in this text.)

Figure 13(a) is a 3-input TRL NOR gate. It is a NOR gate because if any one input is high, or if more than one is high, the output is low. The base resistors R_B are chosen so that the base current, even when only one input is high, is sufficient to drive the transistor into saturation, giving the low output voltage.

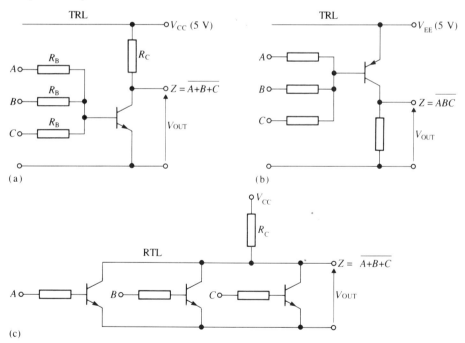

FIGURE 13 (a) An n–p–n TRL circuit with three inputs forming a NOR gate; (b) a p–n–p TRL circuit with three inputs forming a NAND gate; (c) an RTL NOR gate circuit—not to be confused with the TRL circuit

The first point to note is that particular logic families tend to favour the construction of particular types of gates. TRL, using n–p–n transistors, naturally gives rise to NOR gates, as shown. Similarly if a p–n–p transistor were to be used instead of an n–p–n one, the whole circuit would be turned upside down, as in Figure 13(b), and a NAND gate would be produced. (Only if all the inputs are high will the output be low.) This is an example of a general principle, that in complementary circuits (i.e. those which use p–n–p transistors instead of n–p–n ones or p-channel MOSFETs instead of n-channel ones and in which the supply polarities are reversed) a NOR gate is converted to a NAND gate or *vice versa*, and an OR gate is converted to an AND gate or *vice versa*. Gates made from n–p–n bipolars and from n-channel MOSFETs tend to be preferred because they are inherently faster. This implies, for example, that in practice TRL gates are generally NOR gates, that TTL gates are normally NAND gates and that ECL gates normally have both NOR and OR outputs. CMOS however uses both types of MOSFET so this principle doesn't apply.

The main problems that logic circuits are designed to overcome are firstly fan-in and fan-out, secondly noise immunity and thirdly speed of response. The questions of fan-in, fan-out and noise margins—by which noise immunity is specified—are discussed in the next three sub-sections. Switching speeds are considered when particular logic circuits are explained.

3.2 FAN-IN

Fan-in refers to the number of logic inputs to a gate. Commercially available gates are usually made with a specified number of inputs (i.e. a specified fan-in) which may be anywhere between 2 and about 10. (Only inverters have a 'fan-in' of 1.) The reason why there may be limitations to the possible fan-in for particular families of gates can most easily be illustrated by first considering the two simplest kinds of gates: TRL made from bipolar transistors and nMOS made from n-channel MOSFETs. As you will see, the reasons for limitations on fan-in are quite different in the two cases.

The fact that the TRL NOR gate shown in Figure 13(a) has three inputs, means that it is a circuit with a fan-in of 3. The limitation on the possible fan-in soon emerges if we consider the situation where only one input is high—causing the output to be low.

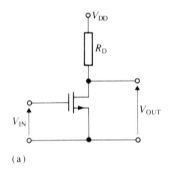

In this circuit, if the high input voltage is equal to V_{CC} and the low input voltage is 0 V, the input base current to the transistor, $I_{B(ON)}$, is evidently

$$I_{B(ON)} = \frac{V_{CC} - V_{BE}}{R_B} - 2 \times \frac{V_{BE}}{R_B}.$$

That is, the current from the high input terminal has to supply both the transistor base current as well as the current taken away by the other two input resistors. With a fan-in of n, the above equation becomes

$$I_{B(ON)} = \frac{V_{CC} - V_{BE}}{R_B} - (n-1) \times \frac{V_{BE}}{R_B}. \tag{14}$$

If $V_{CC} = 5$ V and $V_{BE} = 0.7$ V for the saturated transistor, it is easy to calculate that $I_{B(ON)} \approx 0$ if $n = 7$, which means that the circuit would not work if it had a fan-in of 7. the transistor would not be turned on, let alone driven into saturation. A fan-in of about 5 is all that can be achieved with TRL. As you will see later, with TTL and ECL the limitations of fan-in are largely overcome. A fan-in of about 10 is typical for these families of circuits.

MOSFET circuits may also have fan-in limitations, but for quite different reasons.

Figure 14(a) shows again an n-channel MOSFET inverter made from an enhancement-mode device and a load resistor. Since MOSFETs do not require a d.c. base current, gates cannot be constructed in the same way as those made from bipolars. Instead, to form a NAND gate MOSFETs are put in series as shown in Figure 14(b). In this circuit, only if all three inputs are high will transistors T1, T2 and T3 be conducting and the circuit

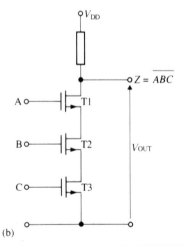

FIGURE 14 (a) A simple MOSFET inverter with a resistor as load. (b) The basic construction of a MOSFET NAND gate

output be low. Hence the circuit is a NAND gate. The problem is that this output voltage may not be low enough to drive the next stage properly if more than two or three MOSFETs are put in series.

You will recall from SAQ 3 that the low logic voltage obtainable from an n-channel MOSFET which has a resistor as a load is about 1 V. This voltage is to be the input voltage for the next stage of a digital circuit, and must therefore be less than the threshold voltage of the next transistor. With three MOSFETs in series, as in Figure 14(b) the output voltage will be about 3 V which is sure to be too high. It may even be too high with only two MOSFETs in series. Thus there is again a limitation on the possible fan-in.

Now, although it is possible to design MOSFET NAND gates which give a better performance than this—such a family of gates is called nMOS—the fact remains that the fan-in of such gates is severely limited. nMOS gates are not available in isolated packages, but they are still widely used in integrated circuits since they occupy very little silicon surface area. Even there, however, nMOS NAND gates are rarely made with fan-in of more than two.

3.3 FAN-OUT

Fan-out is the number of inputs of similar gates that the output of a single gate can drive. With bipolar transistors the fan-out depends partly on the d.c. conditions—in much the same way that fan-in does—and partly on the switching speed expected of the circuit. With MOSFETs speed is the main consideration: fan-out and speed are almost inversely proportional to each other, because the larger the number of input capacitances that a given output current has to drive the slower the rise and fall times are going to be. So, for one reason or the other it is usual to specify a maximum fan-out.

Again, the reason for limitations on fan-out with bipolar transistors can easily be illustrated with TRL. Figure 15(a) shows a 3-input TRL NOR gate whose output is connected to the inputs of five following gates. It is assumed that the NOR gate will provide an output of nearly V_{CC} when the transistor in the gate is cut off. In practice however the effective resistance of all five load resistors in parallel can be sufficient to reduce the output voltage of the NOR gate to a value that is insufficient to drive the following gates. If this were the case a smaller fan-out than 5 would have to be specified for the circuit. Again better circuit designs overcome this problem.

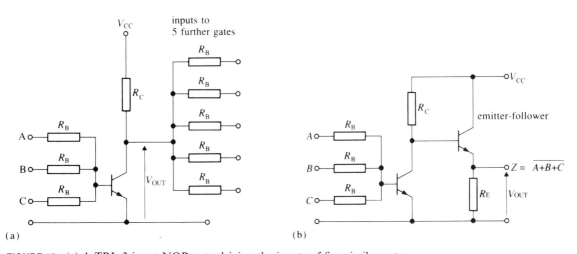

FIGURE 15 (a) A TRL 3-input NOR gate driving the inputs of five similar gates. (b) The addition of an emitter-follower T2 to give the circuit an active pull-up and better fan-out

The kind of output shown in Figure 15(a), in which the current to drive the next input 'high' is supplied through a collector *resistor*, when the transistor is cut off, is called a **passive pull-up**. It is to be contrasted with an **active pull-up** in which an emitter-follower or 'buffer' is added to the NOR gate in place of the resistor, as in Figure 15(b). The emitter-follower in Figure 15(b) reduces the current through R_C by a factor of β and therefore makes a much larger fan-out possible.

The circuit of Figure 15(a) however has an **active pull-down**; that is, a saturated transistor is used to establish the *low* output voltage. There is usually little difficulty in providing the transistor with sufficient base current to ensure that a low output will be established even when a large number of gates are connected to its output.

As a general rule, it is to be expected that active pull-up and active pull-down circuits will be needed if large fan-outs are to be achieved.

Gates with a passive pull-up have one advantage over those with an active pull-up: they can be wired together to make a **wired-AND** gate. Figure 16(a) shows the circuit of wired-AND gate consisting of three 2-input NOR gates which have a common load resistor R_L. In this arrangement each NOR gate has an 'open collector', such as that shown for TRL in Figure 16(b), so that when its output transistor is cut-off it is virtually disconnected from the output of the whole circuit. This means that the output of the wired-AND gate goes low if any of the six inputs is high. Figure 16(a) shows the overall logic function achieved by the circuit, namely the simple extension of the NOR function. The circuit is called a 'wired-AND' gate, even though its overall function is that of a NOR gate, because the wiring together of the *outputs* of the constituent gates produces an AND function. Thus if the outputs of the constituent gates are called P, Q and R, as shown, and the overall output is Z, then $Z = PQR$. The output Z is high only if all the outputs P, Q, R are high.

☐   *Revision*   In the circuit of Figure 16(a) use Boolean algebra to show that the output $Z = PQR$ leads to the overall function $\bar{Z} = A + B + C + D + E + F$.

■   From the figure you can see that $Z = PQR = (\overline{A + B})(\overline{C + D})(\overline{E + F})$, so, by de Morgan's theorem, $\bar{Z} = A + B + C + D + E + F$.

In the same way, logic gates which have passive pull-down outputs (such as p–n–p versions of the TRL circuit) can be combined to form wired-OR gates.

The advantages of the wired gate is that it dispenses with the need for an additional logic gate (either an AND gate or an OR gate) to combine the outputs of several gates.

Such circuits are widely used in computer construction. Several parts of the computer may need to send or receive the same 16-bit words, so instead of interconnecting the several parts of the computer with separate wires, they are all interconnected via a **data bus**, as it is called. The wire interconnecting the gates in Figure 16(a) is a simple illustration of such a data bus. When the transistor in a gate is cut-off it is in effect disconnected from the bus; so that if it is arranged that only one gate output at a time is *not* cut-off, the gates can take it in turns to use the same wire, or 'bus', to send voltages to other parts of the computer, without interfering with each other.

3.4   NOISE MARGINS

As explained in Block 3, Part 2, the voltage levels between which digital circuits switch can be any two values, though in practice they are specified as two *ranges* of values. For example, if the intended *output* voltages of a logic circuit are 5 V and 0 V, the specified input logic levels that the circuits following them can rely upon, may be specified as $\geqslant 4$ V and $\leqslant 1$ V, allowing a 1 V variation from the intended values, for whatever reason. This 1 V

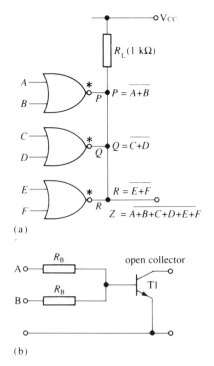

FIGURE 16 (a) Three NOR gates wired together to form a 'wired-AND' gate. (b) An 'open collector' TRL NOR gate for connection into a wired-AND gate. (Open-collector gates are usually indicated in logic diagrams by an asterisk, as in (a).)

variation is called the **noise margin**, NM. Circuits have to be designed to operate with any input voltage levels that lie within the noise margins. The noise margins at the high and low logic levels are not necessarily the same, so it is sometimes necessary to refer to them as the high and low noise margins NM_H and NM_L. However, the high and low noise margins of the standard TTL NAND gates described in Section 3.5.1 are quoted as $NM_H = 0.4$ V and $NM_L = 0.4$ V, as explained in Block 3, Part 2. The larger the noise margins of a circuit family the better, since it means that the circuit is less likely to fail if voltage levels drift away from their intended values or if unwanted voltages are picked up.

The calculations in this text ignore the need to allow for noise margins in circuit design. Taking noise levels into account, as well as considering the effects of parameter variations of circuit components, adds a further level of complexity in circuit design which is beyond the scope of this course. You are therefore only expected to know what noise margins are, not to incorporate them into your calculations of switching times, etc.

3.5 SWITCHING CIRCUITS USING BIPOLAR TRANSISTORS

3.5.1 TTL (TRANSISTOR–TRANSISTOR LOGIC)

A detailed analysis of TTL along the same lines as the analysis of TRL is beyond the scope of this course. So TTL is only described here rather than explained in detail.

The basic circuit from which TTL evolved is shown in Figure 17(a). It is rather like the TRL circuit, but with transistors replacing the input resistors. It is a NAND gate because when an input is high it cuts off the emitter of the input transistor. Thus when both inputs are high all the current through R_B flows into the base of T2 turning it ON and driving it into saturation. When either input is low the current through R_B flows through the corresponding input transistor, starving T2 of current, thus turning it OFF. Therefore the output is low only when both inputs are high—which is the NAND function. The advantage of this arrangement over TRL is that the current drawn from T2 by either T1A or T1B when T2 is being turned off is much greater than in TRL. In fact it is about β times the current in R_B, ensuring that Q_B and Q_{BS} in T2 are removed much more quickly.

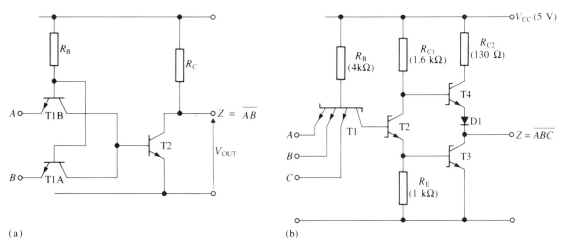

FIGURE 17 (a) The basic circuit from which TTL was evolved; (b) the basic circuit of Schottky-clamped TTL NAND gate with a fan-in of 3

The basic circuit of Figure 17(a) contains several design weaknesses, so the circuit has been developed and improved in several ways, as illustrated in Figure 17(b).

(a) Instead of providing a transistor for each input to the circuit, all the inputs are combined into one multiple-emitter transistor, labelled T1 in Figure 17(b). This transistor is constructed simply by forming a number (up to 10) of emitter p–n junctions opposite one collector junction as shown in Figure 18. So a fan-in of 10 is a typical maximum figure.

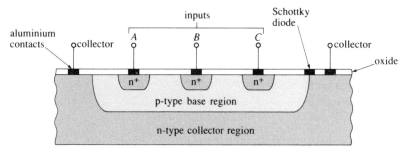

FIGURE 18 A diagrammatic cross-section of the multiple-emitter input transistor of a TTL NAND gate. The diagram also shows how the Schottky diode is connected across the collector-base p–n junction

(b) A 'totem-pole' output stage (as it is called), comprising T3, D1 and T4, is added to the output to give the circuit a good pull-up and pull-down performance capable of driving up to 10 gates (i.e. a fan-out of 10). T4 pulls the output up when T2 and T3 are OFF, and T3 pulls the output down when T2 is turned ON. The diode D1 in the emitter lead of T4 (voltage drop of about 0.7 V) ensures that T4 is not conducting when T3 is conducting by reducing V_{BE} of T4 to much less than 0.65 V. The 130 Ω resistor is for protection; it ensures that T4 is not destroyed by overheating if the output is inadvertently shorted to ground. It has the effect of lowering the high output voltage depending on the specified maximum fan-out.

(c) Schottky diodes are placed across each collector—base junction, as indicated in Figure 17(b) by the special transistor symbols introduced in Figure 9(b). The diodes are formed simply by placing an aluminium contact across the collector p–n junction, as indicated in Figure 18. The metal forms a Schottky diode with the n-type collector region and makes ohmic contact with the p-type base region; so with one relatively simple, additional production process a Schottky diode can be added to each transistor in the circuit. With this addition to TTL the design is sometimes referred to as 'Schottky-clamped' TTL or Schottky TTL or STTL or TTL(S). The penalty paid for the inclusion of these Schottky diodes is that the low output voltage is increased somewhat. The low output voltage becomes about 0.4 V instead of 0.2 V because T3 is not driven so far into saturation with the Schottky diode connected across it.

TTL gates designed as in Figure 17(b) dissipate about 20 mW each. In more recent designs the resistor values are increased to give a lower power version of Schottky-clamped TTL that dissipates only 2 mW per gate. This is called LSTTL—the first 'L' meaning 'low power'. 'Advanced LSTTL' is a further version which dissipates only 1 mW per gate, without sacrificing too much in speed.

SAQ 10 How would you modify the TTL NAND gate circuit of Figure 17(b) so that it was suitable for interconnection as a wired gate? If two such 3-input TTL NAND gates formed a wired gate what would its overall logic function be in terms of the inputs to the two NAND gates?

3.5.2 ECL (EMITTER-COUPLED LOGIC)

The basic inverter circuit of ECL is the *long-tailed pair* (see Part 2) shown in Figure 19(a). If the two transistors are identical—which is relatively easy to arrange in integrated circuits—and if their base terminals are held at the same voltage, the current in the 'long tail' resistor R_E divides equally

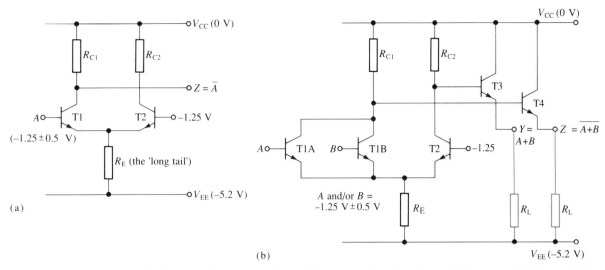

FIGURE 19 (a) The long-tailed pair on which ECL is based. (b) An example of a 2-input ECL gate with NOR and OR outputs

between the two transistors. A change in the differential input voltage by more than a few tenths of a volt will steer most of this current through one or other transistor, causing the output voltage to change. (Remember that a 0.1 V change in V_{BE} produces a current change of about e^4.) So if the input voltage at A *rises* by 0.5 V (whilst the input voltage of B is held constant) most of the current through R_E must pass through T1 causing the output voltage to *fall*. Hence the circuit is an inverter. If the input voltage *falls* by 0.5 V all the current passes through T2 and the output rises to the supply line voltage—usually 0 V as shown. So in ECL the high and low input voltage levels are only about 1 V apart. *The load resistors R_C are therefore chosen so that the output voltage also changes by about 1 V.*

By adding an extra input transistor in parallel with T1, as shown in Figure 19(b), a NOR gate is formed. That is, if either input is high, or if both inputs are high, the collectors of T1A and T1B are low.

The addition of an emitter-follower, T4, whose input is connected to the collector of T1, increases the output current drive from the circuit and therefore increases the possible fan-out. T4 also lowers the high and low output levels by about 0.7 V (i.e. by V_{BE}) so that they are at the right levels for driving subsequent ECL circuits.

A second output, labelled Y, is usually also available, connected to the output of T2 via T3. This gives the complement of Z and so generates an OR function. That is, $Y = A + B$ whilst $Z = \overline{A + B}$. The resistors labelled R_L, and shown in grey in the figure, are not part of the standard ECL circuit; they are included here to represent the loading that the next ECL circuits would in effect provide.

To understand the circuit operation better, consider a particular set of values. Suppose the high and low input voltage levels are 1 V apart: namely -0.75 V for the high level, and -1.75 V for the low level, as indicated in Figure 19(b). If both inputs A and B are initially low, and if the base of T2 is held at -1.25 V, a change from low to high of either input will result in the current through R_E being switched from T2 to pass through one of the input transistors, T1A or T1B.

If R_E is 750 Ω the current to be switched is about $(3 \text{ V}/750 \, \Omega) = 4 \text{ mA}$ (though it changes a bit when the input changes between high and low). To give the required 1 V swing at the output, resistors R_{C1} and R_{C2} must be $(1 \text{ V}/4 \text{ mA}) = 250 \, \Omega$. With these values in the circuit of Figure 19(b) the collector voltage of T2 will rise from -1 V to 0 V as T2 is turned off and the collectors of T1A and T1B will fall from 0 V to -1 V.

Note that when an input is at -0.75 V and the output is at -1 V the collector junction is forward biased by 0.25 V, so the transistor is actually in saturation. But the forward bias of the collector junction is not sufficient to create a significant amount of saturation charge, so it is fair to say that $Q_{BS} \approx 0$.

If the voltage drops across the base-emitter junctions of the emitter-followers T3 and T4 were 0.75 V, the two output voltage levels, at the emitters of T3 and T4, would be brought back to -0.75 V and -1.75 V, as required to drive the next stage. Typical ECL circuits actually give an output voltage swing of less than 1 V, usually between 0.8 and 0.9 V (corresponding to 0.7 V drops across the emitters of T3 and T4), which is all that is needed for the circuit to operate reliably and with reasonable noise margins.

The circuit is extremely fast, giving typical rise and fall times of between 0.5 and 2 ns. This speed is due partly to the transistors never being driven into saturation to any significant extent, partly because the voltage swings are so small—implying short charging times for any load capacitors or internal transistor capacitances—and partly because there is plenty of current available to move charge from one transistor to another. The penalty for achieving this speed is a power dissipation per gate of about 50 mW. This includes the additional fixed-current emitter follower circuit (not shown) needed to hold the base of T2 at -1.25 V.

3.5.3 THE SWITCHING SPEED OF ECL

The switching speed can be estimated in much the same way as for TRL, except that $Q_{BS} = 0$. Consider the inverter circuit of Figure 19(a) when input A goes positive from -1.75 V to -0.75 V. Charge is transferred from transistor T2 to transistor T1. The charges involved are as follows:

(a) A forward bias of 0.5 V is transferred from T2 to T1 involving a transference of charge from one emitter capacitance C_e to the other. This charge is equal to $(C_e \times 0.5$ V);

(b) the reverse bias of the collector-base junction of T2 increases by 1 V as that of T1 decreases by 1 V. This involves a transfer of charge equal to (the collector-base capacitance, $C_c) \times (1$ V);

(c) the base charge $Q_B(\approx \tau_t I_C)$ is transferred from T2 to T1.

So, assuming that the capacitances are linear and do not change with applied voltage, which is only approximately true:

$$Q_{ON} = (C_e \times 0.5 \text{ V}) + (C_c \times 1 \text{ V}) + \tau_t I_C.$$

For the transistor specified in Figure 7 the turn-on charge becomes

$$Q_{ON} \approx (2 \text{ pF} \times 0.5 \text{ V}) + (1.5 \text{ pF} \times 1 \text{ V}) + (0.4 \text{ ns} \times 4 \text{ mA})$$

$$= 4.1 \text{ pC}.$$

The turn-off charge Q_{OFF} (i.e. the charge transferred back again to T2 when T1 is turned off) is the same as Q_{ON}.

In this circuit the *base* current is not the factor limiting speed as it was in TRL because the bases are connected to emitter-followers which produce plenty of current. The current which achieves the transfer of charge from one transistor to the other is that flowing in the resistor R_E, which in this case is about 4 mA. So:

$$t_{ON} \text{ and } t_{OFF} \approx \frac{4.1 \text{ pC}}{4 \text{ mA}} \approx 1.0 \text{ ns}.$$

 SAQ 11 Typical values of t_{ON} and t_{OFF} for commercially available ECL circuits are quoted as 0.5 ns. How do you suppose this reduced value is achieved? Illustrate your answer with relevant calculations.

The same calculation applies to multiple-input gates, except that if there are n inputs (a fan-in of n) there are more capacitances to be charged and discharged by the same current. So fan-in is limited by its effect on the switching times. Similar considerations limit the fan-out; the output of each circuit is not a perfect voltage source so that if it drives too many following gates simultaneously the current available to drive the next circuit will become the factor that limits the switching speed. Typically the maximum fan-out specified is 10.

SAQ 12 Consider only the d.c. operation of the ECL circuit of Figure 19(b). If $R_E = 750\ \Omega$ and $R_{C1} = R_{C2} = 250\ \Omega$ and the β of each transistor is 200:

(a) estimate how much input current is required to hold one of the inputs high.

(b) How much current is available at the output to hold an input high when the noise margin is 0.3 V (i.e. if the output is not to fall by more than 0.3 V due to the load)?

3.6 SWITCHING CIRCUITS USING MOSFETs

There are two kinds of MOSFET logic circuits in widespread current use. The first is called nMOS because it contains only n-channel MOSFETs. The second is CMOS, which uses pairs of transistors, one of which is n-channel and the other is p-channel. CMOS is used in two quite different forms: (a) as standard CMOS; (b) in 'pass gates' or 'transmission gates'. Each of these is described in the next subsections.

3.6.1 nMOS (n-CHANNEL-MOSFET LOGIC)

The circuit diagrams of two forms of nMOS inverters are shown in Figures 20(a) and 20(b). In Figure 20(a) the resistor in the drain lead shown in Figure 11(a) is replaced by a *depletion-mode* n-channel device with its gate connected to its own source so that $V_{GS} = 0$. With this connection a depletion-mode device behaves like a resistor, but a rather non-linear one and is called a **loadMOST**.

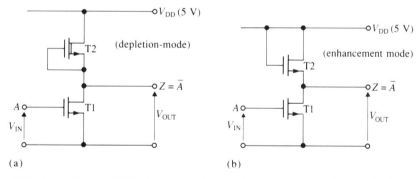

FIGURE 20 (a) An nMOS inverter with a depletion-mode transistor as its load. (b) An nMOS inverter with an enhancement-mode transistor as its load

☐   Figure 21(a) (overleaf) shows the d.c. characteristics of an n-channel depletion-mode MOSFET. Which curve is the d.c. characteristic of this device when it is being used as a loadMOST?

■   A loadMOST has its gate connected to its source so that $V_{GS} = 0$. Therefore the curve labelled $V_{GS} = 0$ is the nonlinear characteristic of this loadMOST.

Figure 20(b) shows an *enhancement-mode* device with its gate connected to its drain replacing the drain resistor. This too behaves like a nonlinear resistor and is therefore an alternative form of loadMOST. Of the two versions shown in the figure, Figure 20(a) gives a rather better performance whilst that in Figure 20(b) tends to be cheaper, because it is simpler to make two enhancement-mode devices on one chip than to make two different devices side by side.

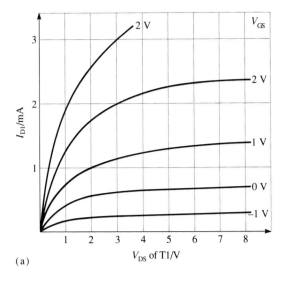

(a)

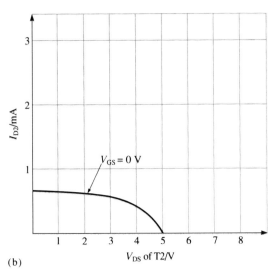

(b)

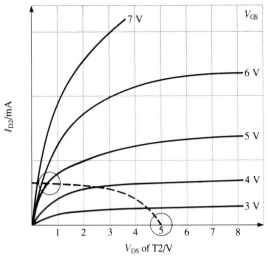

(c)

FIGURE 21 (a) The d.c. characteristics of a depletion-mode, n-channel MOSFET; (b) the $V_{GS} = 0$ curve in (a) drawn as a load line for the circuit of Figure 20(a); (c) the load line construction applied to the circuit of Figure 20(a), showing the operating points when the circuit is used as a switch

In both diagrams in Figure 20, T1 is an *enhancement-mode* n-channel device, so that when its input is low (i.e. less than its threshold voltage) it is cut off. The operating points of either circuit can be found using the load line construction method described in Section 2.1.

For the circuit of Figure 20(a) the load line is shown in Figure 21(b). It is the $V_{GS} = 0$ curve of transistor T1 (shown in Figure 21(a)) drawn on *axes referring to V_{DS} of transistor T2*, as explained in Section 2.1. Then, superimposing this load line on the d.c. characteristics of T2, as in Figure 21(c), shows the possible operating points of the circuit of Figure 20(a). Evidently, referring to the ringed areas on the figure, if V_{GS} of T2 is less than 2 V the output voltage of the circuit will be close to 5 V; and if V_{GS} is close to 5 V the output voltage will be about 0.5 V, leaving a good noise margin.

The way these inverter circuits can be developed into NAND gates or NOR gates is shown in Figure 22. In Figure 22(a) the output voltage will be low only if both inputs are high, giving the NAND function. In the circuit of Figure 22(b) the output is low if either input is high, giving the NOR function.

The complementary family of gates known as pMOS is made from p-channel devices. These gates are inherently slower than nMOS circuits, and so are little used now. They had their day when the technology of making reliable nMOS had not yet arrived. In the early days the p-channel devices were much more stable than the n-channel ones.

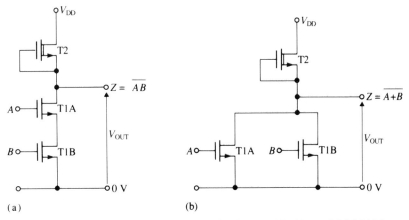

FIGURE 22 (a) An nMOS NAND gate with a fan-in of 2; (b) an nMOS NOR gate with a fan-in of 2

The main rival to nMOS (and pMOS) is CMOS, consisting of complementary pairs of p-channel and n-channel transistors as already explained. Mixing p- and n-channel devices on one chip involves even more processes than mixing enhancement- and depletion-mode devices, but the performance advantages outweigh this disadvantage in many applications.

3.6.2 STANDARD CMOS (COMPLEMENTARY MOSFET LOGIC)

Figure 23(a) shows the basic CMOS inverter. It consists of an n-channel, enhancement-mode MOSFET in series with a p-channel, enhancement-mode one. (Note that the alternative symbols for MOSFETs, which include letters p and n rather than arrows to distinguish between p-channel and n-channel devices, are being used here.)

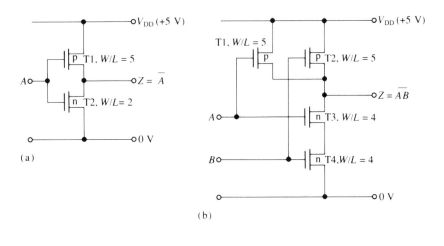

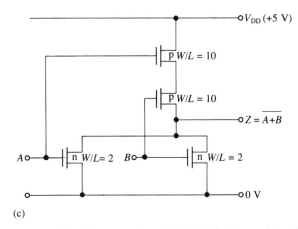

FIGURE 23 (a) A CMOS inverter; (b) a CMOS NAND gate; (c) a CMOS NOR gate. Note the W/L ratios specified for each transistor

In this circuit, when the input is high the p-channel device is cut-off and the n-channel device is conducting. And when the input is low the opposite is true. Since one transistor or the other is always cut-off in the steady state, the current, and therefore power dissipation, are very low. Indeed power dissipation is virtually zero except during the switching process when current flows through one or other transistor. This means that *the power consumption of CMOS is proportional to the number of times per second that the switch operates*. It is this near-zero power consumption, except when switching, which makes CMOS so suitable for digital watches, calculators, etc., in which the gates switch relatively infrequently and are powered by small batteries.

Evidently CMOS has inherently both active pull-up and active pull-down outputs, so large fan-outs are possible. And because one transistor is always in the conducting state, even though it is starved of current by the other transistor, the voltage across it is almost zero. Hence the output voltages are very firmly held at one or other of the supply line voltages.

Complete symmetry of the circuit is not, however, achieved by giving the two transistors the same dimensions. Because the mobility of holes in the p-channel device is less than the mobility of electrons in the n-channel device by a factor of about 2.5, the transconductance of the p-channel device would be correspondingly smaller if the devices had the same channel width and length. So, to compensate for this difference in mobility, the width/length (W/L) ratio of the channel of the p-channel device is usually made about 2.5 times larger than that of the n-channel device, as indicated in the figure. The point of balancing the pull-up and pull-down capability of the circuit is to ensure that the rise and fall times of the circuit are equal.

Another example of the use of a 'tailored' W/L ratio is for the purpose of distributing clock pulses around a computer. One CMOS inverter may have to drive 1000 or more following gates. For such an output stage large W/L ratios may be needed to supply enough current to achieve the required response times.

The switching speed of CMOS is crucially dependent on the channel length in the MOSFETs. This can be understood as follows. The current available at the output of a circuit is a function of the devices' transconductance; which depends on the W/L ratio. Thus for a transistor with a large W/L ratio the drain current per unit gate voltage, I_D/V_{GS}, is larger than for one with a small W/L ratio. The input capacitance of the circuit depends on the area of the gate electrode, which depends on $W \times L$.

But the switching time depends on the ratio $\dfrac{\text{input capacitance}}{\text{input current}}$, which is proportional to $\dfrac{W \times L}{W/L} = L^2$. Hence the circuit's response times decrease in proportion to the square of the channel length. In integrated digital circuits the channel length is fixed by the smallest controllable dimension in the manufacturing process; this is therefore the main parameter in fixing the propagation delay specified for a particular brand of CMOS.

> **SAQ 13** If the transistor whose characteristics are shown in Figure 2(b) has a W/L ratio of 5, how would the scale of the current axis be changed if W were halved and L were doubled? By how much would the gate-channel capacitance of the transistor be changed?

How can this basic inverter be developed into CMOS logic gates?

To produce a 2-input NAND gate it is necessary to put the two n-channel devices in series, as for nMOS, but to achieve the pull-up property that nMOS lacks a p-channel pull-up transistor is provided for each input n-channel device, as shown in Figure 23(b). In this circuit if both inputs are low, T1 and T2 are conducting whilst T3 and T4 are cut-off, so the output is high. Equally if both inputs are high the output will be low.

□   What happens if only one input is high?

■   Driving input A high enables T3 to conduct, but, since T4 is still cut-off and in series with T3, no current will flow and the output remains high. The fact that T1 becomes cut-off when A goes high simply means that it can be ignored. It is parallel with T2, which remains conducting. So the output remains high if only one input goes high.

Thus the output is low only when both inputs are high, which is the NAND function

In this circuit, further thought has to be given to the W/L ratios of the transistors. The pull-up capability of the circuit depends on either T1 or T2, but the pull-down capability depends on T3 and T4 in series. So W/L for T1 or T2 must be 2.5 times $W/(L3 + L4)$ as indicated by the W/L ratios in the figure.

The number of inputs (i.e. the fan-in) can be increased by adding pairs of transistors in a manner which should be clear from this two-input example. The W/L ratios have to be further modified however, as already explained.

Because of the very low input currents involved there would be hardly any limit on the fan-out capability of the circuit if speed did not matter. However, the more inputs of following circuits there are to be driven, the greater the total capacitance and the longer the charging time for a given current drive. So the rise times and fall times of the gates increase almost in proportion to the fan-out.

A 2-input NOR gate in CMOS is shown in Figure 23(c), together with the appropriate W/L ratios. These ratios are not so economical in silicon as they are for NAND gates, given that channel length is made as small as possible in both kinds of gate, so in general NAND gates are to be preferred in CMOS.

SAQ 14	Explain why the NAND gate W/L ratios are more economical in silicon area than those of the NOR gate when minimum sized gates are being designed. Assume the smallest channel length L possible is 5 μm.

3.6.3 CMOS TRANSMISSION GATES

Transmission gates are also known as 'pass gates' and 'linear gates' and 'bi-directional gates'. Their function is to switch current on and off in either direction through them. They can therefore also switch a.c. currents such as speech signals on telephone lines. CMOS transmission gates simply consist of a complementary pair of enhancement-mode MOSFETs in parallel as shown in Figure 24(a). (Note that the transistor symbols with letters instead of arrows are appropriate here because current can flow in either direction through the transistors.) Typically the threshold voltages of the transistors are -1 V for the p-channel device, and $+1$ V for the n-channel device. The transmission gate is switched on by driving the gate G_1 of the n-channel device high at the same time as driving the gate G_2 of the p-channel transistor low. In other words, current can flow through the transmission gate when $G_1 = 1$ and $G_2 = 0$, and is cut-off when $G_1 = 0$ and $G_2 = 1$. Thus if $G_1 = X$ and $G_2 = \bar{X}$, current will flow when $X = 1$ and will be cut-off when $X = 0$. This behaviour is represented by Figure 24(b). The gate can be thought of simply as an electronically operated switch.

It is evident from Figure 24(b) that the speed at which the output terminal of the switch is brought to the logic level (either high or low) of input A, when $X = 1$, depends on both the input capacitance C_{in} of the following circuit and the current passing through the gate. When $X = 0$ input A is disconnected from the load, so the voltage across the load is unaffected when input A changes.

Transmission gates are widely used in *multiplexers*, which means that they are used to connect any one of several inputs to a given load. They differ

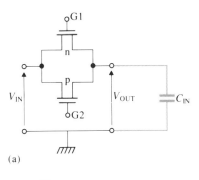

(a)

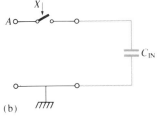

(b)

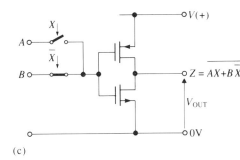

(c)

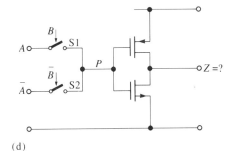

(d)

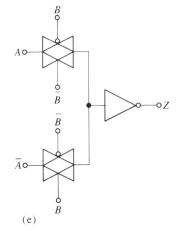

(e)

FIGURE 24 (a) A CMOS transmission gate, (b) a simple representation of a transmission gate; (c) two transmission gates connected to a CMOS inverter to form a 2-input multiplexer; (d) see SAQ 15; (e) the same circuit as (d), but with each transmission gate represented by the standard symbol for it

from OR gates in that they can connect a *signal* through to the load, whose voltage can be any value within limits, not just one of two logic levels. Figure 24(c) shows the simplest of multiplexers, a 2-input multiplexer, connected to a CMOS inverter. Input A is connected to the inverter if $X = 1$ whilst input B is connected to the inverter if $X = 0$. Hence the output of the circuit is the inverse of AX or $B\bar{X}$. That is $\bar{Z} = AX + B\bar{X}$.

Note that if the inputs are analogue signals and the inverter is an analogue inverter made from depletion-mode devices, the circuit is an analogue multiplexer, connecting one or other analogue signal (after inversion) to the output.

SAQ 15	(a) Two transmission gates are connecting the inputs A and $\bar{A}$ to the input of an inverter as shown in Figure 24(d). The gates are controlled as labelled on the diagram. What is the logic function of this gate?

(b)   How would you make a 4-input multiplexer?

Figure 24(e) shows the same circuit as that of Figure 24(d), but redrawn to show the usual symbols for transmission gates.

All kinds of logic gates can be constructed from transmission gates. They have the advantage that it is not necessary to modify the W/L ratios to produce different kinds of gates and different fan-ins.

4   SUMMARY OF PART 3

1   Digital circuits are normally circuits whose output voltages switch between two, fairly well defined levels in response to one or more inputs whose voltages also change between the same two levels. The key characteristics of digital circuits are (a) the speed with which they switch from one level to another (i.e. the *response times*); (b) the number of inputs they have (i.e. their *fan-in*); (c) the number of similar circuits they can drive (i.e. their *fan-out*) and (d) the permitted variation of the output voltage levels of gates, etc, (i.e. the *noise margins*).

2   The *response times* of transistors are specified in terms of the *turn-on time* t_{on} and the *turn-off time* t_{off}. t_{on} can be subdivided into the *delay time* t_d and the *fall time* t_f, and t_{off} for bipolar transistors can be subdivided into the *saturation time* t_s and the *rise time* t_r, and for MOSFETs is just the rise time t_r. Turn-on and turn-off times refer to 10% and 90% changes of the output, respectively, in response to abrupt changes of input voltage.

3   Response times depend on (a) the charge required to turn on or turn off the device namely Q_{ON} and Q_{OFF}, and (b) on the rate at which charge can be supplied to or removed from the input (i.e. the input current). $t_{on} \approx Q_{ON}/I_{ON(av)}$ and $t_{off} \approx Q_{OFF}/I_{OFF(av)}$. Q_{ON} and Q_{OFF} consist of the charge needed to charge or discharge the capacitances in the device, plus any charge stored in the device that has to be supplied or removed.

4   In bipolar transistors charge is stored in the base region when the transistor is conducting. In normal operation the charge is $Q_B = I_C \tau_t$ (where τ_t is the *transit time*; a typical value is 0.5 ns). In saturation an additional charge, the *saturation base charge* Q_{BS} is also stored in the base region. $Q_{BS} = \tau_s(I_B - I_C/\beta)$, where τ_s is the *saturation time constant*. (τ_s is equal to the lifetime of the carriers that form Q_{BS}; a typical value is 40 ns.) Q_{BS} can be greatly reduced by connecting *Schottky diodes* in parallel with the collector-base junctions of bipolar transistors.

5   In MOSFETs negligible charge is stored in the channel and there is no d.c. gate current, so the input gate current has only to charge or discharge the input capacitances, plus any stray capacitances to ground caused by the wiring. Any capacitive loading of the output of a gate slows down the change of output voltage relative to the change of output current.

6   The transistor parameters needed to calculate response times are:

(a) for bipolars: *current gain* β, *transit time* τ_t, *emitter* and *collector junction capacitances*, and, if the transistor is driven into saturation, the *saturation time constant* τ_s;

(b) for MOSFETs: the *gate-source*, *gate-channel* and *gate-drain capacitances* plus the *gain factor* β, and the ratio of the *channel width W* and the *channel length L*.

7   The response times of gates are usually expressed in terms of *propagation delay times* which refer to the delay between the instant of 50% change in the input voltage and the 50% change of the output voltage. $t_{PHL} \approx t_{on}$ and $t_{PLH} \approx t_{off}$.

8   *Active pull-up* and *pull-down* output circuits of gates (rather than *passive* ones) are used to reduce switching times and to increase *fan-out*.

9   The main families of bipolar switching circuits are TTL and ECL. ECL is the fastest, but it dissipates the most power. There are various versions of TTL designed either for high speed or for low power dissipation. The main families of MOSFET switching circuits are nMOS (comprising only n-channel MOSFETs) and CMOS (comprising *complementary pairs* of MOSFETs). With *standard* CMOS, normal gates, such as NAND gates etc., can be formed. With CMOS *transmission gates*, multiplexers (both digital and analogue) as well as normal gates can be constructed. The properties of these main families (except transmission gates which are rather different) are summarized in Table 1.

Table 1 Comparison between logic families

Logic family	nMOS	CMOS	STTL	TTL(LS)	ECL
high/low †(V)	5/1.5	5/0	4/0.3	4/0.5	−0.8/−1.6
supply (V)	5	5	5	5	−5.2
power dissipation per gate (mW)	0.5	0.1*	10	1	50
propagation delay per gate (ns)	10	15	2	4	0.5
fan out	20	50	10	10	25
gates/mm²	400	300	50	50	20

*With CMOS, the power consumption is proportional to the rate at which the gate is switched. The figure given is for a 1 MHz switching rate.
†High/low volts refers to the nominal voltages used to represent logic 1s and logic 0s.

10   Logic circuits with passive pull-up or passive pull-down outputs can be connected to form *wired-AND* or *wired-OR* circuits by giving them a common load resistor.

5 CONCLUDING REMARKS

Inevitably, in an introductory text such as this, the descriptions and explanations presented concentrate on only the main factors which determine the performance of both the devices and the digital circuits which depend upon them. There are, however, various additional factors which usually only have a second-order effect on the properties of circuits, but which in any particular circuit design may have to be taken into account. They need not concern you in this course, beyond being aware that there are such qualifications.

(i) The current gain β in bipolars is not completely independent of the d.c. current flowing through the device. Typically β doubles as I_C increases from microamps to milliamps, and then falls away again at high currents.

(ii) The p–n junction transition-region capacitance is nonlinear: a fact which somewhat modifies all the foregoing estimations of response times. It decreases as the *reverse* bias voltage is increased, and increases when the *forward* bias is increased. This makes it possible to make voltage dependent capacitors and so, for example, to adjust the resonant frequency of *LC* circuits by varying the applied d.c. voltage (see Block 9, Part 2).

(iii) There are limitations to the voltage and current that can safely be applied to transistors, and therefore also to logic circuits. These are specified in data sheets under the heading 'Ratings'. Ratings are maximum values for reliable operation and so should not be exceeded in any good circuit design.

This concludes Block 4. Its purpose has been to explain—without too much analysis—how transistors work and how the basic circuits made from them behave. More detailed explanations of the circuits described, and descriptions of the many other transistor circuits that can be bought nowadays, are to be found in the books listed in the references.

SAQ 1

Figure 25 shows two ringed areas on the d.c. characteristics of a bipolar transistor. One represents the ON state when the current is 2 mA and the output voltage V_{CE} is small, and the other represents the OFF state when the current is almost zero and the output voltage is about equal to the supply voltage of 6 V.

The base currents required are: 30 μA or more when the transistor is ON; and about 2 μA or less when the transistor is OFF.

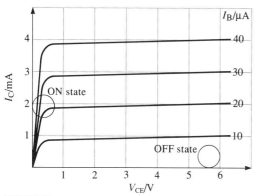

FIGURE 25

SAQ 2

(i) From the intersection of the $I_B = 10$ μA transistor characteristic and the load line you can read off the operating point as $V_{CE} \approx 4$ V and $I_C \approx 0.95$ mA.

(ii) Similarly for $I_B = 40$ μA, the operating point is $V_{CE} \approx 0.2$ V and $I_C \approx 2.9$ mA.

(iii) Since $I_C \approx 0.95$ mA when $I_B = 10$ μA, $\beta = I_C/I_B \approx 95$. Note that I_C increases somewhat as V_{CE} increases—due to the Early effect—so β also increases somewhat.

(iv) The load line needed to achieve the operating points of $I_C = 2$ mA and almost zero is a line from $V_{CE} = 6$ V with a slope of $-1/3$ kΩ, so the required load resistance is 3 kΩ.

SAQ 3

Figure 26 shows the load line corresponding to a load resistance of 4 kΩ. Evidently if the output voltage is to be

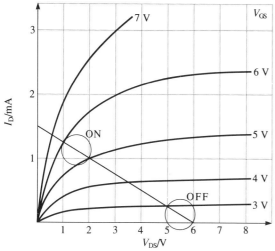

FIGURE 26

greater than 5 V the input voltage V_{GS} must be less than 3 V, as indicated by the circled area labelled OFF. And if the input is to be between 5 V and 6 V the output voltage must lie within the area labelled ON, between about 1 V and 2 V. This is a suitable voltage for driving the next stage into the OFF condition. These two output voltages (i.e. greater than 5 V and between 1 V and 2 V) are compatible with the required input voltages (i.e. less than 3 V and between 5 V and 6 V).

SAQ 4

The load line you should have drawn on Figure 1(b), corresponding to a load resistance of 3 kΩ, is a straight line from $V_{CE} = 6$ V to $I_C = 2$ mA. So:

(i) From Figure 1(b), with this new load line, $V_{CE(sat)} \approx 0.1$ V.

(ii) Similarly $I_C \approx 1.95$ mA.

(iii) V_{BE} is still about 0.65 V, even though V_{CE} is only 0.1 V, so

$$R_B = \frac{6 \text{ V} - 0.65 \text{ V}}{40 \text{ μA}} = 134 \text{ kΩ}.$$

(iv) the ratio $I_C/I_B = \dfrac{1.95 \text{ mA}}{40 \text{ μA}} \approx 49$. (Note that this is much less than β.)

SAQ 5

(i) The increase of emitter current, as compared with the collector current, is represented by the greater gradient of the minority carrier density profile at the emitter end of the base region. (A curved minority carrier density profile—such as that shown in Figure 4(b)—always accompanies the presence of significant recombination in the base region.)

(ii) The forward bias of the collector junction is represented by the fact that the minority carrier density at the collector end of the base region is greater than the equilibrium density (see Part 1 of this block).

Since the minority carrier density at the emitter end of the base region has increased in saturation, the forward bias V_{BE} of the emitter junction must have increased too.

SAQ 6

Here you have to repeat the calculations of the worked example, but with a new value for R_B.

(a) First calculate $I_{B(ON)}$ and $I_{B(OFF)}$.

With $R_B = 30$ kΩ it follows that

$$I_{B(ON)} = (5 \text{ V} - 0.65 \text{ V})/30 \text{ kΩ} = 145 \text{ μA},$$

and that $I_{B(OFF)} = 0.65 \text{ V}/30 \text{ kΩ} = 22$ μA.

(b) Next calculate Q_B and Q_{BS}.

$I_{C(ON)} = 4.8$ mA as in the worked example, so $Q_B = 1.92$ pC, but

$$Q_{BS} = \tau_s(I_{B(ON)} - I_C/\beta)$$

$$= 20 \text{ ns} \times (145 \text{ μA} - 4.8 \text{ mA}/200) = 2.42 \text{ pC}.$$

(c) So the approximate response times for this circuit are

Delay time (equation (5))

$$t_d \approx \frac{(C_e + C_c) \times 0.65 \text{ V}}{I_{B(ON)}} = \frac{3.5 \text{ pF} \times 0.65 \text{ V}}{145 \text{ μA}} = 15.7 \text{ ns.}$$

Fall time (equation (6))

$$t_f \approx \frac{Q_B + C_c \times \Delta V_{CE}}{I_{B(ON)}} = \frac{1.92 \text{ pC} + 1.5 \text{ pF} \times 4.8 \text{ V}}{145 \text{ μA}} = 63 \text{ ns.}$$

Saturation time (equation (7))

$$t_s \approx \frac{Q_{BS}}{I_{B(OFF)}} = \frac{2.42 \text{ pC}}{22 \text{ μA}} = 110 \text{ ns.}$$

Rise time (equation (8))

$$t_r \approx \frac{Q_B + C_c \times \Delta V_{CE}}{I_{B(OFF)}} = \frac{1.92 \text{ pC} + 1.5 \text{ pF} \times 4.8 \text{ V}}{22 \text{ μA}}$$

$$= 415 \text{ ns.}$$

Hence (equation (9)) $t_{on} \approx t_d + t_f \approx 79$ ns,

and (equation (10)) $t_{off} \approx t_s + t_r \approx 525$ ns.

Notice how t_s has increased whilst the other times have decreased.

SAQ 7

(i) In the circuit of Figure 7(a) we know that $I_{B(OFF)} = 0.65$ V/50 kΩ = 13 μA, that $I_{C(ON)} \approx 4.8$ mA and that $I_{B(ON)} = 4.35$ V/50 kΩ = 87 μA.

We also know from equation (7) that $t_s \approx \dfrac{Q_{BS}}{I_{B(OFF)}}$

and that, from equation (2), $Q_{BS} = \tau_s(I_{B(ON)} - I_{C(ON)}/\beta)$.

So substituting for Q_{BS} in the equation (7) for t_s gives

$$t_s = \frac{\tau_s(I_{B(ON)} - I_{C(ON)}/\beta)}{I_{B(OFF)}} = \frac{\tau_s(87 \text{ μA} - 24 \text{ μA})}{13 \text{ μA}}$$

so $\tau_s = \dfrac{t_s \times 13 \text{ μA}}{63 \text{ μA}} = \dfrac{50 \text{ ns} \times 13 \text{ μA}}{63 \text{ μA}} = 10.3$ ns.

SAQ 8

Substituting the given values in the diode equation (neglecting the -1 in the brackets) gives

$$10^{-3} \text{ A} = 10^{-9} \text{ A} \times \exp(V_D/0.025 \text{ V})$$

or

$$\ln 10^6 = V_D/0.025$$

so

$$V_D = 13.8 \times 0.025 = 0.345 \text{ V.}$$

SAQ 9

Assume that the gate capacitance is divided in the ratio 2:1 so that in effect $C_{gs} = 0.6$ pF and $C_{gd} = 1$ pF.

(i) *Delay time*: No drain current flows until the threshold voltage has been reached, therefore

$$t_d = \frac{(C_{gs} + C_{gd}) \times V_T}{I_{(ON)}} = \frac{1.6 \text{ pF} \times 1.5 \text{ V}}{400 \text{ μA}} \approx 6 \text{ ns.}$$

(ii) *Fall time*: During turn-on the voltage across C_{gs} increases by a further 3.5 V as V_{GS} rises to 5 V, whilst the voltage across C_{gd} changes by $\Delta V_{GS} + \Delta V_{GD} = 3.5$ V + 4 V = 7.5 V as V_{DS} falls to 1 V. Therefore

$$t_f \approx \frac{C_{gs} \times 3.5 \text{ V} + C_{gd} \times 7.5 \text{ V}}{I_{(ON)}}$$

$$= \frac{0.6 \text{ pF} \times 3.5 \text{ V} + 1 \text{ pF} \times 7.5 \text{ V}}{400 \text{ μA}} = 24 \text{ ns.}$$

(iii) *Rise time*: The voltage changes across the capacitances are of the same magnitude as for the fall time. The drain current has fallen to zero when V_{GS} has fallen to the threshold voltage. Therefore

$$t_r = \frac{0.6 \text{ pF} \times 3.5 \text{ V} + 1 \text{ pF} \times 7.5 \text{ V}}{400 \text{ μA}} = 24 \text{ ns.}$$

SAQ 10

Referring to the circuit of Figure 17(b), the circuit modification needed is the creation of an open-collector output. Hence it is only necessary to remove T4, R_{C2} and D1, then T3 has an open collector when the output transistor is cut off.

The wired gate described is shown in Figure 27. If the inputs to the two NAND gates are A, B and C and D, E and F and their outputs are P and Q, respectively, as shown, then the overall output $Z = PQ$.

Therefore $Z = (\overline{ABC})(\overline{DEF})$.

By de Morgan's theorem $\bar{Z} = ABC + DEF$, so $Z = \overline{ABC + DEF}$.

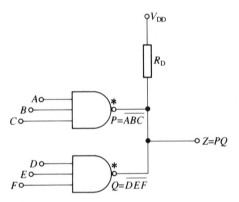

FIGURE 27

SAQ 11

As explained in the text, voltage swings of about 0.8 V between logic levels, rather than 1 V, are typical, so that the charge required by the capacitances is 2 pC rather than 2.5 pC.

Not much is gained by decreasing R_E, and thus increasing the emitter currents, because Q_B increases in proportion. For example, if I_E were increased to 5 mA, Q_B would be 2 pC instead of 1.6 pC, giving a total Q_{ON} of $(2 + 2)$ pC and a turn-off time of 0.8 ns.

Evidently it is necessary to use transistors with smaller capacitances and/or a smaller transit time. With $C_e = 1.5$ pF, $C_c = 0.8$ pF and $\tau_t = 200$ ps, $Q_{ON} = (1.5 \text{ pF} \times 0.4 \text{ V}) + (0.8 \text{ pF} \times 0.8 \text{ V}) + (200 \text{ ps} \times 4 \text{ mA}) = 2.04$ pC

So $t_{ON} \approx \dfrac{2.04 \text{ pC}}{4 \text{ mA}} \approx 0.5$ ns.

SAQ 12

(a) To hold an input transistor of the ECL circuit conducting a base current of $I_C/\beta(\text{min})$ is required. Since either I_{C1} or I_{C2} is 4 mA, depending on which is conducting, and the minimum β is 200, the base current needed is $= 4 \text{ mA}/200 = 20$ μA.

(b) The output of the emitter-follower T3 (say), when T2 is cut off, is capable of delivering a current of β times its base current. The noise margin allowed is 0.3 V, which is the maximum voltage allowable across R_{C2}, so the base current to T3 is (0.3 V/250 Ω) = 1.2 mA. Therefore the output current from the emitter follower, if the minimum β is 200, is about 240 mA.

Note this allows a very large fan-out, provided switching speed is not a significant consideration.

SAQ 13

Since the current axis depends on the W/L ratio, the current scale in Figure 2(b) would be changed by a factor of 4 to give the characteristics of the revised transistor. That is the 1 mA calibration would become 0.25 mA. The gate-channel capacitance would not however be affected by the change in the channel dimensions, since $W \times L$ is not altered.

SAQ 14

The area ($= W \times L$) occupied by the channels of the 4 devices in the NAND gate is

Area $= 2 \times (20\,\mu m \times 5\,\mu m) + 2 \times (25\,\mu m \times 5\,\mu m)$

$= 450$ square microns.

For the NOR gates the area comes out to be

Area $= 2 \times (10\,\mu m \times 5\,\mu m) + 2 \times (50\,\mu m \times 5\,\mu m)$

$= 600$ square microns.

(Note: A micron is a millionth of a meter.)

SAQ 15

(a) First work out the logic level of node P. It is convenient to tabulate the circuit performance in a truth table, as follows:

A	B	P	Output Z (Note that $\bar{Z} = P$)
0	0	1	0
0	1	0	1
1	0	0	1
1	1	1	0

(Note: Switch S2 is closed (i.e. $\bar{B} = 1$) and $\bar{A} = 1$.)

The circuit is therefore an EXCLUSIVE-OR gate

(b) A 4-input multiplexer is like Figure 24(c), but with 4 input transmission-gate switches, only one of which is closed at any one time. The switches must therefore be controlled by two logic levels, say X and Y. Then the four switches are controlled by the outputs of four AND gates (or their equivalent) whose outputs are XY, $\bar{X}Y$, $X\bar{Y}$, $\bar{X}\bar{Y}$.

REFERENCES

1   Hodges, D. A. and Jackson, H. G. (1983) *Analysis and Design of Digital Integrated Circuits*, McGraw Hill.
2   Horowitz, P. and Hill, W. (1980) *The Art of Electronics*. Cambridge

INDEX OF KEY TERMS